Lecture Notes in Mathematics

A collection of informal reports and seminars
Edited by A. Dold, Heidelberg and B. Eckmann, Zürich

Series: Institut de Mathématique, Université de Strasbourg
Adviser: M. Karoubi and P. A. Meyer

256

Carlos A. Berenstein
Harvard University, Cambridge, MA/USA

Milos A. Dostal
Stevens Institute of Technology, Hoboken NJ/USA

Analytically Uniform Spaces and their Applications to Convolution Equations

Springer-Verlag
Berlin · Heidelberg · New York 1972

AMS Subject Classifications (1970): 42 A 68, 42 A 96, 35 E 99, 46 F 05

ISBN 3-540-05746-3 Springer-Verlag Berlin · Heidelberg · New York
ISBN 0-387-05746-3 Springer-Verlag New York · Heidelberg · Berlin

Offsetdruck: Julius Beltz, Hemsbach/Bergstr.

Preface

About twelve years ago Leon Ehrenpreis created a theory which culminated in what he called the fundamental principle for linear partial differential equations with constant coefficients.* This theory yields practically all results about PDE's and systems of PDE's as corollaries of a new Fourier type representation for their solutions. The possibility of such a representation is essentially the content of the fundamental principle. However the whole subject remained unpublished until recently, when two special monographs appeared, both giving complete proofs as well as a great number of far reaching applications.** Nevertheless, in view of the amazing complexity of the whole material, neither of these monographs enables the reader to penetrate rapidly into the heart of the subject. The main purpose of the present volume is to provide such an introduction to this beautiful field which represents a promising area for further research. In order to achieve this goal, the authors decided to treat only the case of one PDE. Indeed, all the basic ideas appear in this case, while one can still avoid building the huge machinery which is necessary for the proof of the general statement (cf. the first four chapters in either of the aforementioned monographs). In proving the main theorem (see Chapter IV below), the authors adopted the original approach of Ehrenpreis. However in the concrete presentation and choice of the material they mainly followed their previous publications.

* Abbreviated in the sequel as PDE's.

**V. P. Palamodov, "Linear differential operators with constant coefficients," Moscow 1967; L. Ehrenpreis, "Fourier analysis in several complex variables," Wiley-Interscience 1970. The latter monograph contains many applications going far beyond PDE's.

Some of the results appear here for the first time (for a more detailed account, see the section of bibliographical remarks at the end of this volume).

Let us now characterize very briefly the contents of these lecture notes. The main idea of the subject consists in a systematic use of Fourier transforms in the study of convolution operators acting on different function spaces. However, instead of dealing directly with concrete function spaces and their duals, one considers a large class of spaces satisfying certain natural conditions (the class of analytically uniform spaces). The definition and basic properties of these spaces can be found in Chapter I. Chapter II is devoted to one important family of analytically uniform spaces, namely the class of Beurling spaces. This chapter also serves as an illustration of the difficulties one has to overcome in proving that a given space enters the scheme defined in Chapter I. Another class of analytically uniform spaces is treated in Chapter III. Moreover this chapter contains an application of these spaces to certain convolution equations. The basic ideas of Chapter III can also serve as motivation for Chapter IV, where the fundamental principle is established. The concluding section contains the bibliographical remarks as well as some further comments concerning the results discussed in the text.*

The present lecture notes were originally based on a course given by the second author at the University of Strasbourg (Strasbourg, France) in the Spring of 1970; and, on a similar course given later by the first author at Harvard University. However in its final form the text differs rather substantially from both of these courses.

*Formulas, definitions, theorems, etc., are numbered throughout each chapter separately. Thus, for instance, "Lemma 3.II" refers to Lemma 3 in Chapter II, while "Theorem 2" means Theorem 2 of the same chapter in which the reference appears; raised numbers such as [2] refer to the section of the bibliographical remarks.

It is a great pleasure for the authors to express their sincere thanks to Professor P.-A. Meyer of the University of Strasbourg for his kind invitation to publish this volume in the Springer series, "Lecture Notes in Mathematics"; and, to Professor Leon Ehrenpreis of Yeshiva University in New York for his constant support and interest. The authors also extend warm appreciation to Mr. William Curley and Dr. Marvin Tretkoff for carefully checking the manuscript; and, to Miss Mary Jo Vogelsang and Miss Connie Engle for their excellent typing job.

C.A.B., M.A.D. *

* The first author was supported by the U. S. Army Office of Research (Durham); the second author wishes to thank "Institut de recherches mathématiques avancées" in Strasbourg (France) for various forms of support.

Contents

CHAPTER

CHAPTER I

Definition and Basic Properties of Analytically Uniform Spaces

§1. INTRODUCTION

About fifty years ago a new trend appeared in mathematical analysis, and since then it has been flourishing until the present day. To characterize its earlier period, it suffices to recall the names of S. Bochner, K. O. Friedrichs, J. Hadamard, F. John, I. Petrovskij, M. Riesz, S. Sobolev, and N. Wiener. Their work significantly changed such classical areas of mathematics as the theory of partial differential equations and Fourier analysis.

The next stage of this development was marked by the appearance of the celebrated treatise of L. Schwartz, "Théorie des distributions" (1950-51; cf. [46]). The importance of this work for analysis was twofold. First of all, the classical concept of a function was broadened by introducing more general objccts called distributions (or generalized functions) on which the standard operations of analysis can easily be defined. The second and equally important achievement of this theory was the extension of Fourier analysis to certain classes of these generalized functions. As a consequence, the classical theory of Fourier series and integrals became applicable to many functions which are large at infinity (more exactly, to generalized functions of polynomial growth, cf. [46][1]). Since then the theory of distributions combined with complex variable techniques developed into a powerful tool in mathematical analysis.

Once the proper framework had been found it became possible to formulate properly, and later to solve, many of the basic problems of PDE's.[2] The pioneering work in this field is connected above all with the names of L. Ehrenpreis, L. Hörmander, and B. Malgrange. To give a better idea about the type of problems we have in mind, we shall briefly discuss one of them.

First, let us recall some basic definitions and the corresponding notation. Let Ω be a non-empty open subset of $\mathbb{R}^n$. A sequence $\{K_s\}_{s\geq 1}$ is said to exhaust Ω (notation: $K_s \nearrow \Omega$), if all K_s are non-empty compact subsets of Ω, $K_s \subset \operatorname{int} K_s$ $(s \geq 1)$ and $\bigcup_{s\geq 1} K_s = \Omega$. For each $s \geq 1$, $\mathcal{D}(K_s)$ is defined as the space of all C^∞-functions with support in K_s. $\mathcal{D}(K_s)$ is endowed with the Fréchet topology of uniform convergence of functions and their derivatives on the set K_s. Then we define

$$(1) \qquad \mathcal{D}(\Omega) = \lim_{s\to\infty} \operatorname{ind} \mathcal{D}(K_s)^{*}$$

Hence $\mathcal{D}(\Omega)$ is an $(\mathcal{LF})$-space *; and, it is not difficult to see that the definition of $\mathcal{D}(\Omega)$ does not depend on the particular choice of the exhausting sequence $\{K_s\}$. $\mathcal{D}'(\Omega)$ denotes the dual space to $\mathcal{D}(\Omega)$, i.e. the space of all Schwartz distributions on Ω. If $\alpha = (\alpha_1,\ldots,\alpha_n)$ is any multiindex of nonnegative integers, $|\alpha|$ will denote its length, i.e. $|\alpha| = \alpha_1+\ldots+\alpha_n$; and, for $D = \left(-i\frac{\partial}{\partial x_1},\ldots,-i\frac{\partial}{\partial x_n}\right)$ and any vector $\xi = (\xi_1,\ldots,\xi_n) \in \mathbb{R}^n$, we set

$$(2) \qquad D^\alpha = (-i)^{|\alpha|} \frac{\partial^{|\alpha|}}{\partial x_1^{\alpha_1} \ldots \partial x_n^{\alpha_n}} \; ; \qquad \xi^\alpha = \xi_1^{\alpha} \ldots \xi_n^{\alpha} .$$

If $P(\xi_1,\ldots,\xi_n)$ is a polynomial, $P(D) = P(D_1,\ldots,D_n)$ denotes the corresponding partial differential operator.

*By this we shall always mean a strong inductive limit of Fréchet spaces.

Now the problem can be formulated as follows:

Given an arbitrary P as above, when is the equation

(3) $$P(D)u = f$$

solvable in $\mathcal{D}'(\Omega)$ for any $f \in \mathcal{D}'(\Omega)$?

At a first glance, this would seem to be a typical problem in the functional analysis. In abstract terms one could formulate it in the following way. Let T be a continuous injective mapping of an $(\mathcal{L}\mathcal{F})$-space E into another $(\mathcal{L}\mathcal{F})$-space F. When is the adjoint mapping $T': F' \to E'$ surjective? The Hahn-Banach theorem shows that it suffices to find conditions under which T is an open mapping. (Indeed, to show that $T'u = f$ has a solution u in F' for each $f \in E'$, we first observe that this equation already defines u on the range of T, because for each $\psi = T\phi$, $\langle\psi,u\rangle = \langle\phi,f\rangle$. If the functional $\psi \mapsto \langle\psi,u\rangle$ is continuous on TE, it can be extended to the whole space F; but the continuity of $\psi \mapsto \langle\psi,u\rangle$ follows from the continuity of T^{-1}, i.e. the openness of T.) However, since the mapping T generally is not surjective, the openness of T cannot be proved by standard methods. We are thus led to the following problem: Given an injective continuous mapping $T: E \to F$, where E and F are $(\mathcal{L}\mathcal{F})$-spaces, find necessary and sufficient conditions for T to be open. However this turns out to be a difficult problem, a fully satisfactory solution of which has not yet been found.[3]

All of this indicates that in order to solve equations such as equation (3), one has to combine functional analysis with yet different methods. The above-mentioned problem concerning equation (3) was completely solved by Hörmander [26,28]. Combining functional analysis with Fourier transforms, Hörmander found necessary and sufficient conditions for the solvability of equation (3) in $\mathcal{D}'(\Omega)$. On the other hand, the most systematic use of Fourier analysis in this field was made by L. Ehrenpreis who found a unified way of studying different problems of the above type by Fourier transform.* Before concluding this section, let us sketch very briefly the motivation which underlies Ehrenpreis's approach.

Given any $\phi \in \mathcal{D}(\Omega)$ we define the Fourier transform of ϕ by

$$(4) \qquad \mathcal{F}(\phi)(\zeta) = \hat{\phi}(\zeta) = \int_{\mathbb{R}^n} e^{-i\langle x,\zeta\rangle}\phi(x)\,dx$$

$$(\zeta = \xi + i\eta \in \mathbb{C}^n).$$

The space of all functions $\hat{\phi}$, $\phi \in \mathcal{D}(\Omega)$, will be denoted by $\hat{\mathcal{D}}(\Omega)$. The topology $\hat{\mathcal{T}}(\Omega)$ on $\hat{\mathcal{D}}(\Omega)$ is defined by requiring that $\mathcal{F}:\mathcal{D}(\Omega) \to \hat{\mathcal{D}}(\Omega)$ be an isomorphism of locally convex spaces. One important consequence of formula (4) is that the elements of $\hat{\mathcal{D}}(\Omega)$ are entire functions, i.e. $\hat{\mathcal{D}}(\Omega)$ is a subspace of the space $\mathcal{A}$ of all entire functions in $\mathbb{C}^n$. The space $\mathcal{A}$ will always be considered in the topology of uniform convergence on compact subsets in $\mathbb{C}^n$; therefore, $\hat{\mathcal{D}}(\Omega)$ is continuously embedded in $\mathcal{A}$. Then, however, problems of type (3) can be viewed as problems of mappings between various subspaces of $\mathcal{A}$ carrying a finer topology than the relative topology of $\mathcal{A}$. Since in all concrete situations these subspaces are characterized by different types of growth conditions, a good way of describing their topology seems to be the following:

Let us call a majorant any positive continuous function on $\mathbb{C}^n$. If $\mathcal{K} = \{k\}$ is some non-empty family of majorants, we set

*This method also has its limitations; see Example 3 below and the Bibliographical remarks.

$$(5)\begin{cases} \mathcal{A}(\mathcal{K}) = \left\{ f \in \mathcal{A} : \|f\|_k \overset{\text{def}}{=} \sup\limits_{\zeta \in \mathbb{C}^n} \dfrac{|f(\zeta)|}{k(\zeta)} < \infty \quad (\forall k \in \mathcal{K}) \right\} \\ \\ \mathcal{V}(k;\varepsilon) = \{ f \in \mathcal{A}(\mathcal{K}) : \|f\|_k \leq \varepsilon \} \; ; \quad \mathcal{V}(k) = \mathcal{V}(k;1) . \end{cases}$$

Naturally, for some $\mathcal{K}$ we may obtain $\mathcal{A}(\mathcal{K}) = \{0\}$. The space $\mathcal{A}(\mathcal{K})$ is equipped with the topology $\mathcal{T}(\mathcal{K})$ generated by the norms $\|\cdot\|_k$, $k \in \mathcal{K}$; $\mathcal{A}(\mathcal{K})$ then becomes a Hausdorff locally convex space. Since each k is a positive continuous function on $\mathbb{C}^n$, the space $\mathcal{A}(\mathcal{K})$ is continuously embedded in $\mathcal{A}$. In particular, each $\mathcal{A}(\mathcal{K})$ is a complete space.

Now it is natural to pose the following

<u>Problem</u>. Assume that $(E,\mathcal{T}_E)$ is a given space of entire functions, which is continuously embedded in $\mathcal{A}$. Is it possible to represent the space $(E,\mathcal{T}_E)$ in the form $(\mathcal{A}(\mathcal{K}),\mathcal{T}(\mathcal{K}))$? In other words, given a concrete space $(E,\mathcal{T}_E)$ with the above properties, the question is whether there exists a family of majorants $\mathcal{K}$ such that

$$\text{(A)} \qquad E = \mathcal{A}(\mathcal{K})$$

$$\text{(B)} \qquad \mathcal{T}_E = \mathcal{T}(\mathcal{K}) .$$

If such a family $\mathcal{K}$ exists, we shall call the space $\mathcal{A}(\mathcal{K})$ a <u>complex representation</u> of the space E.

Some explanatory remarks should clarify the foregoing problem. First, the problem should not be understood as the question of whether <u>each</u> space, which can be continuously embedded into $\mathcal{A}$, allows a representation with properties (A) and (B) (cf. Example 3 below). Instead, the problem consists of finding complex representations of concrete spaces which are important in applications. Moreover, even if we already know that some space has a complex representation we may ask for another one; namely, we can look for a new family $\mathcal{K}$ which satisfies some other conditions in addition to (A) and (B) (cf. [4, 17, 23]). Obviously, $\mathcal{K}$ is not uniquely determined by conditions (A) and (B). It is also easy to see that conditions (A) and (B) are independent.

Ehrenpreis was first to recognize the importance of such complex representations for solving linear equations of convolution type in various spaces of distributions. He also found complex representations for most of the known function spaces [20, 21, 23]. The present lectures are intended as an introduction to these topics.

§2. GENERAL PROPERTIES OF ANALYTICALLY UNIFORM SPACES

In the sequel all topological vector spaces are always assumed to be Hausdorff and locally convex. We shall call them l.c. spaces. Given an l.c. space E, let E_b' be the strong dual of E and $<.,.>_E$ the bilinear form defining the duality between E and E_b'. For all other terminology and facts related to l.c. spaces, cf. [24, 32, 43].

Definition 1. An l.c. space W is called an analytically uniform space (AU-space) of dimension n provided the following conditions are satisfied:

(i) W is the strong dual of some l.c. space $(U,\mathcal{T}_U)$.

(ii) There exists a continuous analytic embedding ω of the n-dimensional complex space $\mathbb{C}^n$ into W such that the range $\omega(\mathbb{C}^n)$ is a total subset of W. In particular, for each $S \in U$,

$$\hat{S}(z) = <S,\omega(z)>_U$$

is an entire function in $\mathbb{C}^n$. The mapping $\mathcal{F}: S \to \hat{S}$ is obviously linear and injective. Let $\hat{U} \overset{\text{def}}{=} \{\hat{S}: S \in U\}$, $\mathcal{T}_{\hat{U}} = \mathcal{F}(\mathcal{T}_U)$. Hence $\mathcal{F}: U \to \hat{U}$ is an isomorphism of l.c. spaces.

(iii) There exists a family $\mathcal{K} = \{k\}$ of majorants (cf. §1), which defines a complex representation of $\hat{U}$, i.e. $\mathcal{K}$ such that

(A) $\mathcal{A}(\mathcal{K}) = \hat{U}$,

(B) $\mathcal{T}(\mathcal{K}) = \mathcal{T}_{\hat{U}}$.

Each $\mathcal{K}$ with properties (A) and (B) is called an analytically uniform structure (AU-structure) for the space W. The space U is called the base of the AU-space W.

(iv) There is an AU-structure $\mathcal{K} = \{k\}$ such that if we form for each $N > 0$ the family $\mathcal{K}_N = \{k_N\}$, where

$$k_N(z) \overset{\text{def}}{=} \max_{|z-z'|\leq N} \{k(z')(1 + |z'|)^N\}, \tag{6}$$

then $\mathcal{K}_N$ is again an AU-structure for W.*

(v) There exists a family $\mathcal{M} = \{m\}$ of majorants with the following properties. For each $m \in \mathcal{M}$ and $k \in \mathcal{K}$, $m(z) = \mathcal{O}(k(z))$. Hence all sets

$$A(m,\alpha) \overset{\text{def}}{=} \left\{S \in U: \sup_{z \in \mathbb{C}^n} \frac{|\hat{S}(z)|}{m(z)} < \alpha\right\} \qquad (\alpha > 0;\ m \in \mathcal{M})$$

are bounded in U; moreover, we require that the family $\{A(m;\alpha)\}$ defines a fundamental system of bounded sets in the space U. Each family $\mathcal{M}$ with these properties will be called a bounded analytically uniform structure (BAU-structure) for the space W.

(vi) Similarly as in (iv), all modifications $\mathcal{M}_N$ of some $\mathcal{M}$ are again BAU-structures for W.

Remarks: 1. The families $\mathcal{K}$ and $\mathcal{M}$ are not uniquely determined. Obviously, the uniqueness could be achieved by requiring that these families be the maximal families with properties described in Definition 1.

*Families $\mathcal{K}_N$ will be called modifications of $\mathcal{K}$.

2. Given $S \in U$, the entire function $\hat{S}$ will be called the Fourier transform of S. This terminology is fully justified, because in all the examples we shall consider, W will be a space of functions or distributions in the variable x, $x \in \mathbb{R}^n$, and ω will be the exponential mapping, $\omega: z \mapsto e^{i<x,z>}$.

It follows from condition (iv) above that multiplication by a polynomial defines a continuous endomorphism of $\hat{U}$. In the case of a distribution space W, multiplication by a polynomial P(z) in $\hat{U}$ corresponds to the partial differential operator P(D) acting on W. This suggests the following definition.

Definition 2. If F is an entire function such that the multiplication by F is a continuous endomorphism in $\hat{U}$, i.e. $F \in L(\hat{U},\hat{U})$, F is called a multiplier of the space U; and, the continuous operator $\mathcal{C}_F$ defined as the adjoint of the mapping $S \to \mathcal{F}^{-1}(F(\mathcal{F}(S)))$ is called a convolutor of the space W. The set $\mathcal{C}(W)$ of all convolutors of W is a subspace of L(W,W); and, $\mathcal{C}(W)$ is given the corresponding relative (compact open) topology of L(W,W). Each convolutor $\mathcal{C}_F$ such that the corresponding multiplication by F defines an open endomorphism of $\hat{U}$, will be called invertible in W, and the multiplier F slowly decreasing in $\hat{U}$.

Remark 3. One can verify as in §1 that invertibility of $\mathcal{C}_F$ implies $\mathcal{C}_F(W) = W$. The terminology "slowly decreasing" comes from the following condition which is sufficient for a convolutor $\mathcal{C}_F$ to be invertible in every AU-space W:

There exist positive constants A and N such that for all $z \in \mathbb{C}^n$ there is a ρ_z, $0 < \rho_z \leq N$, such that

(7) $$\min_{|z-z'|=\rho_z} |F(z')| \geq \frac{A}{(1+|z|)^N} .$$

Indeed, given $k \in \mathcal{K}$, condition (iv) shows that, for each $N > 0$, there is a $\tilde{k} \in \mathcal{K}$ and $C > 0$, such that $\mathcal{V}(\tilde{k}_N) \subset \mathcal{V}(k;C)$ (cf. (5)). If $F\hat{S} \in \mathcal{V}(\tilde{k})$, inequality (7) implies

(8) $$|\hat{S}(z)| \leq A^{-1} \max_{|z'| \leq N} (|F(z+z')\hat{S}(z+z')|(1+|z|)^N) \leq A^{-1}\tilde{k}_N(z) ,$$

i.e. $\hat{S} \in \mathcal{V}(k;C/A)$; hence the multiplication by F is an open mapping of $\hat{U}$.[4]

Since condition (iv) of Definition 1 implies that all polynomials are convolutors for any AU-space W, one can ask which polynomials are invertible for a given space W. The answer is simple:

Proposition 1. Every polynomial satisfies condition (7). Therefore all equations (3) are solvable in any AU-space W.

The proof of a more precise version of this proposition appears in Chapter IV (cf. Lemma 2, IV).

Remarks: 4. The above definition of a convolutor is correct only if the mapping $F \to \mathcal{C}_F$ is one-to-one. This, however, is an immediate consequence of analyticity of F.

5. Solvability of any equation (3) in some W is sometimes referred to as solvability of the division problem in W. Therefore the impossibility of solving the division problem in some W implies that W is not an AU-space.

We have just seen that each partial differential operator* P(D) defines a homomorphism of any AU-space W. Therefore it is natural to expect that AU-spaces are nuclear. Actually much more is true [5]:

Theorem 1. If W is an AU-space and U its base, then both U and W are nuclear.

In the proof we shall need a simple lemma on entire functions:

Lemma 1. Let $\ell(z)$ be a majorant in $\mathbb{C}^n$ and $H \in \mathcal{A}$. Set $\Delta_1 = \{z = (z_1, z_2, \dots, z_n) : \max_j |z_j| \leq 1\}$; $\tilde{\ell}(z) = \sup_{z'-z \in \Delta_1} \{\ell(z')(1+|z'|)^{2n+1}\}$; and $d\rho(z) = \pi^{-n}(1+|z|)^{-2n-1}|dz|$ where $|dz|$ is the Lebesgue measure in $\mathbb{C}^n = \mathbb{R}^{2n}$. Then

$$\sup_{z \in \mathbb{C}^n} \frac{|H(z)|}{\tilde{\ell}(z)} \leq \int_{\mathbb{C}^n} \frac{|H(z)|}{\ell(z)} d\rho(z) . \tag{9}$$

Proof. The mean value property of harmonic functions implies

$$H(z) = \frac{1}{\pi^n} \int_{\Delta_1} H(z+\zeta)|d\zeta| . \tag{10}$$

Multiplying the integrand in (10) by the function $\tilde{\ell}(z)/(\ell(z')(1+|z'|)^{2n+1})$, which is ≥ 1 in the polydisk Δ_1, we obtain

$$|H(z)| \leq \frac{1}{\pi^n} \int_{\Delta_1} |H(z+\zeta)||d\zeta| \leq \frac{\tilde{\ell}(z)}{\pi^n} \int_{\mathbb{C}^n} \frac{|H(\zeta)||d\zeta|}{\ell(\zeta)(1+|\zeta|)^{2n+1}} ,$$

and this proves the lemma.

*i.e. P linear and with constant coefficients as we shall always assume.

The proof of Theorem 1 is based on the following criterion for nuclearity of a strong dual E' of an l.c. space E (cf. [43], Proposition 4.1.6). If A is a bounded, closed and absolutely convex subset of E, let E(A) be the normed space $E(A) \overset{\text{def.}}{=} \bigcup_{\lambda \geq 0} \lambda A$ with the norm p_A defined by the unit ball A. The unit ball in the dual space E'(A) will be denoted by A^o. The set A^o is a compact space in the weak topology $\sigma(E'(A), E(A))$; and, we have a natural embedding $\iota: E(A) \to C(A^o)$ where $C(A^o)$ denotes the space of continuous functions on A^o. Let $|\iota|$ denote the mapping of E(A) into $C(A^o)$ defined by $|\iota|(x)$ = the absolute value of the function $\iota(x)$.

<u>Nuclearity of a strong dual</u>:

The space E' is nuclear if and only if E has a fundamental system $\mathcal{N}(E)$ of bounded closed absolutely convex sets such that for each $A \in \mathcal{N}(E)$ there exists a $B \in \mathcal{N}(E)$ and a positive Radon measure μ on A^o for which $\lambda A \subset B$, for some $\lambda > 0$; and, for each $x \in E(A)$,

$$(11) \qquad p_B(x) \leq \langle |\iota|(x), \mu \rangle_{C(A^o)}$$

(Obviously, we can also write $\langle |\iota|(x), \mu \rangle_{C(A^o)} = \int_{A^o} |\langle x, a \rangle_E| \, d\mu(a)$.)

In our case E = U and $\mathcal{N}(E)$ will be defined by means of any family $\mathcal{M}$ (satisfying conditions (v), (vi) of Def. 1) as follows: Let us set $t(\mathcal{M}) = \mathcal{M}_{2n+1} = \{m_{2n+1}\}_{m \in \mathcal{M}}$, $t^2(\mathcal{M}) = t(t(\mathcal{M})), \ldots, \tilde{\mathcal{M}} = \bigcup_{n \geq 1} t^n(\mathcal{M})$; then $\mathcal{N}(E)$ is the family of all sets $A(m, \alpha)$ (cf. Def. 1), $m \in \tilde{\mathcal{M}}$.

Let $A = A(m, \alpha)$ be a fixed set in $\mathcal{N}(E)$. For each $z \in \mathbb{C}^n$, let $\gamma(z)$ be the functional defined on $S \in E(A)$ by $\langle S, \gamma(z) \rangle = \hat{S}(z)/m(z)$. Then γ maps $\mathbb{C}^n$ continuously into A^o. For each $s \in A^o$, let $\delta(s)$ be the element of $C'(A^o)$ defined by $\langle f, \delta(s) \rangle_{C(A^o)} = f(s)$ for all $f \in C(A^o)$. Let us now consider the continuous mapping

$\delta\circ\gamma$: $\mathbb{C}^n \to C'(A^o)$. Integrating this mapping with respect to the measure $d\rho(z)$ of Lemma 1, we obtain a measure $\mu \in C'(A^o)$ such that for the elements of $C(A^o)$ of the form $|f|$, $f \in C(A^o)$, we have

(12) $$\langle |f|,\mu\rangle_{C(A^o)} = \int_{\mathbb{C}^n} |\langle f,\delta(\gamma(z))\rangle_{C(A^o)}|\,d\rho(z)$$

On the other hand, if $S \in E(A)$,

$$\langle \iota(S),\delta(\gamma(z))\rangle_{C(A^o)} = \langle S,\gamma(z)\rangle_{E(A)} = \frac{\hat{S}(z)}{m(z)} ;$$

hence by (12),

(13) $$\langle |\iota|(S),\mu\rangle_{C(A^o)} = \int_{\mathbb{C}^n} \frac{|\hat{S}(z)|}{m(z)}\,d\rho(z) .$$

Let $B = A(m_{2n+1};\alpha)$. Then $B \in \mathcal{N}(E)$ and inequality (9) holds by Lemma 1. However comparing (9) and (13) we obtain (11), which proves the nuclearity of W. The proof of the nuclearity of U is similar [5], and uses a criterion which is "dual" to the one above [43].

The next corollary expresses some well-known properties of nuclear spaces (cf. [43]).

Corollary 1. Let W,U be as above. Let us consider on the space $\hat{U}$ (in addition to the norms $\|\cdot\|_k$, $k \in \mathcal{K}$) the following systems of norms

(14) $$\|\hat{S}\|_1^{(k)} = \int_{\mathbb{C}^n} \frac{|\hat{S}(z)|}{k(z)}\,d\rho(z) \qquad (k \in \mathcal{K})$$

(15) $$\|\hat{S}\|_2^{(k)} = \left(\int_{\mathbb{C}^n} \frac{|\hat{S}(z)|\,d\rho(z)}{k^2(z)}\right)^{\frac{1}{2}} \qquad (k \in \mathcal{K})$$

Then each of the systems $\|\cdot\|_k$ $(k \in \mathcal{K})$, (14) and (15) defines the same topology on $\hat{U}$.* In particular, the topology of $\hat{U}$ can be defined by the the scalar products

$$(16) \qquad [\hat{S}_1,\hat{S}_2]_k = \int_{\mathbb{C}^n} \frac{\hat{S}_1(z)\overline{\hat{S}_2(z)}\,d\rho(z)}{k^2(z)} \qquad (k \in \mathcal{K}) .$$

Let us denote by $\hat{W}$ the dual of the space $\hat{U}$. The space $\hat{W}$ corresponds to the space W by the formula $\langle S,T\rangle_U = \langle \hat{S},\hat{T}\rangle_{\hat{U}}$ for any $T \in W$ and $S \in U$. Let $T \in W$ be fixed. By Corollary 1, there exists k, $k \in \mathcal{K}$, such that $\hat{T}$ defines a bounded linear functional on the pre-Hilbert space $(\hat{U},[\cdot,\cdot]_k)$. Let $\hat{U}_{(k)}$ be the completion of this space. The mapping

$$H \mapsto \tilde{H} = \frac{H(z)}{k(z)(1+|z|)^{n+\frac{1}{2}}}$$

is an isomorphism of the space $\hat{U}_{(k)}$ onto a closed subspace $\tilde{U}_{(k)}$ of $L^2(\mathbb{C}^n)$. If $\tilde{T}$ is the image of $\hat{T}$ in this isomorphism, then $\tilde{T}$ can be extended to the functional $\tilde{\tilde{T}}$ defined on the whole space $L^2(\mathbb{C}^n)$ by setting, for instance, $\tilde{\tilde{T}} = 0$ in the orthogonal complement of $\tilde{U}_{(k)}$ in $L^2(\mathbb{C}^n)$. Thus we have $\langle H,\hat{T}\rangle_{\hat{U}} = \langle \tilde{H},\tilde{\tilde{T}}\rangle_{L^2(\mathbb{C}^n)}$. Let $F(z)$ be the function in $L^2(\mathbb{C}^n)$ generating the functional $\tilde{\tilde{T}}$, i.e.

$$(17) \qquad \langle G,\tilde{\tilde{T}}\rangle_{L^2(\mathbb{C}^n)} = \int_{\mathbb{C}^n} G(z)\overline{F(z)}\,|dz|$$

for all $G \in L^2(\mathbb{C}^n)$. Applying this representation to the elements $G \in L^2(\mathbb{C}^n)$ of the form $G(z) = \widetilde{\hat{S}(z)}$, $S \in U$, we obtain

*and thus also on U.

<u>Corollary 2</u>. For any $T \in W$, there exists a majorant $k \in \mathcal{K}$ and a function $F(z) \in L^2(\mathbb{C}^n)$ such that T can be written as the Fourier integral

$$(18) \qquad T = \int_{\mathbb{C}^n} \omega(z) \frac{\overline{F(z)}\,|dz|}{k(z)(1+|z|)^{n+\frac{1}{2}}}$$

The integral in (18) is to be understood in the functional sense.

<u>Remarks</u>: 6. In many examples of AU-spaces the integral in (18) converges as a Lebesgue integral. Moreover, each Fourier representation (18) can be written in the form

$$(19) \qquad T = \int_{\mathbb{C}^n} \omega(z)(1+|z|)^{n+\frac{1}{2}} \frac{d\mu(z)}{k(z)}$$

where μ is the Radon measure in $\mathbb{C}^n$ given by

$$(20) \qquad d\mu(z) = \frac{\overline{F(z)}\,|dz|}{(1+|z|)^{2n+1}} .$$

For $S \in U$, formulas (17) and (19) imply

$$(21) \qquad \langle S,T\rangle_U = \int_{\mathbb{C}^n} \hat{S}(z) \frac{d\mu(z)}{k(z)} .$$

Actually it can be shown (cf. [19]) that for each T, there is a Radon measure μ * and a $k \in \mathcal{K}$ such that (21) holds without assuming (iv), (v), (vi). Then, however, one has to assume that for all $S \in U$ and $k \in \mathcal{K}$, $\hat{S} = \sigma(k)$. Such spaces will be called <u>weak AU-spaces</u>.

*not necessarily of the form (20).

7. Very often it is not necessary to integrate in the integral of (21) over the whole space $\mathbb{C}^n$, but only over a smaller set δ. If δ can be taken the same for all elements $T \in W$, δ is called a __sufficient__ set for W. By the maximum principle, each δ of the form $\delta = \mathbb{C}^n \smallsetminus K$, where K is an arbitrary fixed compact set, is sufficient for every AU-space W. However it frequently happens (cf. [23,51]) that there are sufficient sets of smaller dimension than dim W:

Example 1.   It is shown in [23] that the space $\mathcal{A}$ is itself an AU-space with the base $U_{\mathcal{A}}$ such that $\hat{U}_{\mathcal{A}}$ is the space of all entire functions of exponential type. For the sake of simplicity, let n = 1. Using the Phragmén-Lindelöf principle one can prove [19,23] that the union of any two non-parallel lines in the complex plane is a sufficient set for $\mathcal{A}$.   Moreover, it has recently been proved by B.A. Taylor [51] that the set of all lattice points $\{m + in\}$ (m,n integers) in the plane is a sufficient set for $\mathcal{A}$; in particular, every entire function f can be written as

$$f(z) = \sum_{m,n} a_{m,n} e^{(m+in)z} .$$

8. Another interesting problem of this kind is to find subsets δ in $\mathbb{C}^n$ which would be sufficient for representation of all elements of a given subspace W_0 of W. Given an AU-space of distributions and P(D) a linear partial differential operator with constant coefficients, set $W_0 \overset{\text{def}}{=} \{f \in W : P(D)f = 0\}$. Then the main result on AU-spaces, the so-called __fundamental principle__ (cf. [18,23,41] and Chap. IV below), asserts that the set $V_p = \{\zeta : P(\zeta) = 0\}$ is

(essentially) a sufficient set for W_o.

9. It is clear that if W is an AU-space and A an equicontinuous subset of W, then there exists a majorant $k \in \mathcal{K}$ and a constant $C > 0$ such that the Fourier representation (21) holds with the same k for all $T \in A$; and, for the total variation of $d\mu$ we have $\|d\mu\| \leq C$. This suggests the question whether the converse of this statement is also true. More exactly, let W be a reflexive l.c. space satisfying conditions (i) and (ii). Furthermore, assume that there is a family of majorants $\mathcal{K} = \{k\}$ such that: (iii*) every equicontinuous set can be represented uniformly (with respect to k's) by Fourier integrals of the form (21); (iv*) every Fourier integral of the form (21) is an element of W; and, (v*) for each $k \in \mathcal{K}$, the set

$$\left\{T = \int_{\mathbb{C}^n} \omega(z) \frac{d\mu(z)}{k(z)} : \|d\mu\| \leq 1\right\}$$

is equicontinuous. Is the space $\mathcal{A}(\mathcal{K})$ then a complex representation of $\hat{W}'$? In particular, is W an AU-space? The answer is in general negative, since it suffices to take as U any reflexive proper subspace of a reflexive space U_1 which is an AU-base of some W_1. Then $W = U_b'$ is the counterexample. This indicates that even for concrete spaces W the converse of Corollary 2 may be difficult to prove. In many cases this problem is equivalent to an approximation problem in $\mathcal{A}(\mathcal{K})$ (cf. [23], p. 461-462).

There are some other properties of U and W which follow from Theorem 1.

Corollary 3 Let W be an AU-space with base U. Then, in addition to being nuclear, the spaces U and W always possess the following properties:

(a) U is complete, semireflexive and the bounded sets in U are metrizable and relatively compact (therefore also separable).

(b) W is barreled, separable and the bounded sets in W are precompact. Moreover, if U is barreled,* then (U,W) is a (reflexive) pair of Montel spaces, and the bounded subsets of W are also separable.

(Proof: That U is complete and W separable follows from Def. 1. As a nuclear complete space, U is semireflexive [43]; hence W is barreled [32]. Metrizability of bounded sets in U is proved in [5]. The rest follows easily.)

Remark 10. Actually, the reflexivity of the pair (U,W) was included in the original definition [18,19]. In this case condition (i) can be dropped from Def. 1. However, as the next example shows, there are non-reflexive AU-spaces.

Example 2. Let $|\cdot|$ by any norm in $\mathbb{R}^n$. For each integer $\ell > 0$, set $B_\ell = \{x: |x| \leq \ell\}$. Denote by $\mathcal{D}^s_\ell$, $s=0,1,\ldots$, the space $\{f \in C^s_o(\mathbb{R}^n):$ supp $f \subseteq B_\ell$ with the natural topology of a Banach space,

$$\|f\|_{s,\ell} = \sup_{x;\,|\alpha| \leq s} |D^\alpha f(x)| \;.$$

Let us set

$$\mathcal{D}^s = \lim_{\ell} \text{ind } \mathcal{D}^s_\ell \;, \quad \mathcal{D}_F = \lim \text{proj } \mathcal{D}^s .$$

*In fact, a less restrictive condition is still sufficient [5].

On the other hand we have (cf. §1),

$$\mathscr{D}(B_\ell) = \lim_s \operatorname{proj} \mathscr{D}^s_\ell , \quad \mathscr{D} = \lim_\ell \operatorname{ind} \mathscr{D}(B_\ell) .$$

It is obvious that as sets, $\mathscr{D}_F = \mathscr{D} = C_0^\infty(\mathbb{R}^n)$. The identity mapping $\mathscr{D} \to \mathscr{D}_F$ is continuous [46]. If A is a bounded subset of $\mathscr{D}_F$, then A is bounded in every $\mathscr{D}^s$. Since the spaces $\mathscr{D}^s$ are strict inductive limits, A must be bounded in some $\mathscr{D}^s_{\ell_s}$. In particular, $A \subset \mathscr{D}^0_{\ell_0}$, i.e. all $f \in A$ have the support contained in B_{ℓ_0}. However, since the relative topology of $\mathscr{D}^s$ on $\mathscr{D}^s_{\ell_0}$ coincides with the topology of $\mathscr{D}^s_{\ell_0}$ and A is bounded in $\mathscr{D}^s$, A is also bounded in $\mathscr{D}^s_{\ell_0}$ for all $s \geq 0$; hence A is bounded in $\mathscr{D}(B_\ell)$ and thus also in $\mathscr{D}$. Therefore the bounded sets in $\mathscr{D}$ and $\mathscr{D}_F$ coincide, and the space $\mathscr{D}'_F \overset{\text{def}}{=} (\mathscr{D}_F)'_b$, called <u>the space of distributions of finite order</u>, has the relative topology of the space $\mathscr{D}'$. By [46], $\mathscr{D}'_F$ is dense in $\mathscr{D}'$. Thus $(\mathscr{D}'_F)'_b = (\mathscr{D}')'_b = \mathscr{D}$. As a projective limit of nuclear spaces, $\mathscr{D}_F$ is also nuclear [43]. (Since it can be shown [23] that $\mathscr{D}'_F$ is an AU-space with the base $\mathscr{D}_F$, the nuclearity of both $\mathscr{D}_F$ and $\mathscr{D}'_F$ also follows from Theorem 1.)

We can summarize the properties of $\mathscr{D}_F$ and $\mathscr{D}'_F$ in the following table (cf. Corollary 3):

Property \ Space	$\mathscr{D}_F$	$\mathscr{D}'_F$
nuclear	yes	yes
complete	yes	no
semireflexive	yes	no
reflexive	no	no
barreled	no	yes
bornological	no	yes

The space $\mathscr{D}_F$ is not barreled since it is not reflexive. However, being complete but not barreled, $\mathcal{D}_F$ cannot be bornological [32].* Finally to see that $\mathcal{D}'_F$ is bornological we proceed as follows: $\mathcal{D}'^s$, the strong dual of $\mathcal{D}^s$, is metrizable ($\mathcal{D}'^s = \lim_\ell \text{proj}\ \mathcal{D}'^s(B_\ell)$), and thus bornological. However, since $\mathcal{D}_F$ is dense in each $\mathcal{D}^s$, we have by [24], p. 148, Th. 1.6, $\mathcal{D}'_F = \lim_s \text{ind}\ \mathcal{D}'^s$, and bornologicity of $\mathcal{D}'_F$ follows.

Let W_1, W_2 be AU-spaces of dimensions n_1, n_2 and with bases U_1, U_2 and AU-structures $\mathcal{K}_1$, $\mathcal{K}_2$ respectively. Set $W = W_1 \otimes W_2$, $U = U_1 \otimes U_2$, $\mathcal{K} = \{k_1(z_1)k_2(z_2): k_1 \in \mathcal{K}_1,\ k_2 \in \mathcal{K}_2\}$ and $\omega(z_1,z_2) = \omega_1(z_1) \otimes \omega(z_2)$. Let $\tilde{U}$ be the completion of U in the topology $\mathfrak{T}(\mathcal{K})$. If one of the spaces U_1, U_2 is barreled, then all topologies compatible with the tensor product $U_1 \otimes U_2$ coincide with $\mathfrak{T}(\mathcal{K})$ [5], and the completion $\tilde{W}$ of W in the finest topology on $W_1 \otimes W_2$ (i.e. in the ι-topology of Grothendieck on $W_1 \otimes W_2$, cf. [43]) is an AU-space with the base $\tilde{U}$ and the AU-structure $\mathcal{K}$ (cf. [5]). The space $\tilde{W}$ will be called the <u>AU-product of W_1 and W_2</u>. This remark is useful in various problems involving AU-spaces in several variables, because the possibility of decomposing an AU-space into a tensor product of 1-dimensional AU-spaces very often turns out to be of primary importance. Actually, if, in addition to this tensor property of W, its base U satisfies some further conditions, then the main result of the theory, the so-called fundamental principle, can be established (cf. Chap. IV). Let us summarize these conditions in the following definition.

<u>Definition 3.</u> An AU-space W of dimension <u>n</u> is called a <u>product localizable</u> (or PLAU-) space provided the following holds:

*That $\mathcal{D}_F$ is not bornological also follows directly from the previous discussion of bounded sets in $\mathcal{D}$ and $\mathcal{D}_F$.

(vii) There are 1-dimensional AU-spaces W_j, with U_j, $\mathcal{K}^{(j)}$, $\mathcal{M}^{(j)}$ ($j=1,2,\ldots,n$) as in Def. 1, such that W is the AU-product of the spaces W_j. Moreover, there is a BAU-structure $\mathcal{M}$ of W such that each $m \in \mathcal{M}$ is of the form

$$m(z_1,\ldots,z_n) = m_1(z_1)m_2(z_2)\ldots m_n(z_n) ,$$

where $m_j \in \mathcal{M}^{(j)}$.

The next condition must hold for each W_j, and for this reason we write there, for any fixed $j=1,\ldots,n$, $W, U, \mathcal{K}, \mathcal{M}$ in place of W_j, U_j, $\mathcal{K}^{(j)}$, $\mathcal{M}^{(j)}$ respectively.

(viii) The family $\mathcal{M}$ can be chosen so that, for each $\varepsilon > 0$ and each $m \in \mathcal{M}$, there exists $m^* \in \mathcal{M}$ such that for any $z_0 = x_0 + iy_0 \in \mathbb{C}$, there are entire functions $\phi(z)$ and $\psi(z)$ for which

$$\left.\begin{array}{l} \dfrac{m(z_0)\,|\phi(z)|}{\min\limits_{|\zeta-z_0|\le\varepsilon} |\phi(\zeta)|} \le m^*(z) \\[2ex] \text{and} \\[2ex] |\psi(z)| \sup\limits_{\xi \text{ real}} \left\{ \dfrac{m(\xi+iy_0)}{\min\limits_{|t-y_0|\le\varepsilon} |\psi(\xi+it)|} \right\} \le m^*(z) \end{array}\right\} \text{ for all } z \in \mathbb{C}$$

It follows from the above discussion of AU-spaces that in a certain sense these spaces represent the largest class of l.c. spaces which can be studied by means of the Fourier transform. Nevertheless it is interesting to observe that there exist spaces whose duals can be described by Fourier transforms, but which do not enter the scheme of Def. 1.

Example 3. (The space of real analytic functions on the line.[6])

Given $\varepsilon > 0$ and $K_n = [-n,n]$, let $\mathcal{A}_{\varepsilon,n}$ be the space of all functions continuous on $K_{n,\varepsilon} = \{z \in \mathbb{C}: \mathrm{dist}(z,K_n) \leq \varepsilon\}$ and holomorphic inside $K_{n,\varepsilon}$. $\mathcal{A}_{\varepsilon,n}$ is a Banach space and $\mathcal{R}(K_n) = \lim\limits_{\varepsilon \to 0} \mathrm{ind}\, \mathcal{A}_{\varepsilon,n}$ is a strict inductive limit. For each $m > n$, the natural injection $\mathcal{R}(K_m) \to \mathcal{R}(K_n)$ is a compact mapping and thus $\mathcal{R} = \lim\limits_{n \to \infty} \mathrm{proj}\, \mathcal{R}(K_n)$ is an $(\mathcal{FM})$-space. (cf. [24], p. 109). In particular, $\mathcal{R}$ is reflexive. Obviously, $\mathcal{R}$ is the space of all real analytic functions on the line. We claim that $\mathcal{R}$ is not an AU-space. In the proof we shall need a simple lemma on interpolation.

Lemma 2. Given positive numbers $\varepsilon_n \searrow 0$ and complex numbers z_n such that $\varepsilon_n |z_n| \nearrow \infty$ and $|z_{n+1}| \geq \max(4|z_n|, n^2)$, there exists an entire function $\phi(z)$ such that $\phi(z_n) = e^{\varepsilon_n |z_n|}$ and, for each $\varepsilon > 0$, $|\phi(z)| \leq C_\varepsilon e^{\varepsilon |z|}$ for some $C_\varepsilon > 0$ and all z.

Proof. Let us set

$$f(z) = \prod_{k=1}^{\infty} (1 - \frac{z}{z_n}) .$$

This product is convergent and represents an entire function of order < 1. We claim that

(22) $$|f'(z_n)| > \frac{c}{|z_n|} .$$

Indeed,

$$f'(z_n) = -z_n^{-1} \prod_{k \neq n}^{\infty} (1 - (z_n/z_k)) .$$

For $k < n$, $|1 - (z_n/z_k)| \geq 1$; and, for $k > n$, $|1 - (z_n/z_k)| > 1 - 4^{n-k}$.

Therefore,

$$\left|\prod_{k \neq n}^{\infty} (1 - \frac{z_n}{z_k})\right| \geq \prod_{k > n}^{\infty} \left|1 - \frac{z_n}{z_k}\right|$$

$$\geq \prod_{j=1}^{\infty} (1 - \frac{1}{4^j}) \geq 2 - \sqrt[3]{e} = c$$

$$> 0$$

and (22) follows. Now we define ϕ as

(23) $$\phi(z) = \sum_n \frac{e^{\varepsilon_n |z_n|}}{f'(z_n)} \frac{f(z)}{z - z_n} (\frac{z}{z_n})^{\mu_n}$$

where $\mu_n = [\varepsilon_n |z_n|] + 1$.* First we observe that $\phi(z)/f(z)$ is analytic in $\{z: |z - z_n| > 1 \ (\forall n)\}$. In fact, for such points z,

$$\left|\frac{\phi(z)}{f(z)}\right| \leq \frac{1}{c} \sum_n d_{\mu_n} |z|^{\mu_n} ,$$

where we denoted $d_{\mu_n} = e^{\varepsilon_n |z_n|} |z_n|^{1-\mu_n}$. However the series $h(z) = \Sigma d_{\mu_n} z^{\mu_n}$ is obviously convergent everywhere. This shows that ϕ is entire. For the order ρ_ϕ of the function ϕ we find $\rho_\phi \leq 1$. If $\rho_\phi < 1$, we are done. If $\rho_\phi = 1$, we find that ϕ is of minimal type, and the lemma follows.

Using this lemma we shall prove that there is no family $\mathcal{K}$ for which $\hat{\mathcal{R}}' = \mathcal{A}(\mathcal{K})$. Assume the contrary. Then $\hat{\mathcal{R}}' = \mathcal{A}(\mathcal{K})$ for some $\mathcal{K}$. First we claim that for each $k \in \mathcal{K}$ and $\ell > 0$, there is an

*[a] denotes the integral part of the number a.

$\varepsilon_\ell > 0$ such that

$$\text{(24)} \qquad \exp(\varepsilon_\ell|x| + \ell|y|) = \mathcal{O}(k(z)) \ .$$

If it were not so, one could find sequences $\varepsilon_n \searrow 0$ and $|z_n| \to \infty$ (the latter one growing arbitrarily fast and $y_n = \operatorname{Im} z_n \geq 0$) for which

$$\text{(25)} \qquad \exp(\varepsilon_n|x| + \ell y_n) \geq nk(z_n) \ .$$

Let ϕ be the entire function of Lemma 2 and $F(z) = e^{-i(\ell+\varepsilon_1)z}\phi(z)$. Then, by the Pólya-Martineau theorem [39], $F \in \hat{\mathcal{R}}'(K_{\ell'})$, for some $\ell' > \ell+\varepsilon_1$, whence $F \in \hat{\mathcal{R}}'$. (Let us recall that $\mathcal{R}'$ is an inductive limit of $\mathcal{R}'(K_\ell)$, because $\mathcal{R}$ is dense in each $\mathcal{R}(K_\ell)$; cf. [24], p. 143.) Therefore, $|F(z)| \leq Ck(z)$ with some $C > 0$. However, by (23), (25)

$$|F(z_n)| = e^{(\ell+\varepsilon_1)y_n + \varepsilon_n|z_n|} \geq nk(z_n) \ ,$$

which is a contradiction.

Next we claim that (24) holds with some ε independent of ℓ. Let $\ell \geq 1$ be arbitrary but fixed. Then

$$\text{(26)} \qquad \begin{cases} \frac{\varepsilon_1}{3}|x| + \frac{\ell}{3}|y| \leq \ell|y| \leq \varepsilon_\ell|x| + \ell|y| & \ldots \text{ if } \varepsilon_1|x| \leq 2\ell|y|; \\[2ex] \frac{\varepsilon_1}{3}|x| + \frac{\ell}{3}|y| \leq \frac{\varepsilon_1}{6}|x| + \frac{\varepsilon_1}{3}|x| \leq \varepsilon_1|x| + |y| & \ldots \text{ if } \varepsilon_1|x| \geq 2\ell|y|. \end{cases}$$

Thus, for each ℓ and z, we have

$$\exp(\tfrac{\varepsilon_1}{3}|x| + \tfrac{\ell}{3}|y|) = \mathcal{O}(k(z)) \ .$$

By Corollary 2 (cf. Remark 6), every real analytic function can be written as a Fourier integral,

(27) $$h(s) = \int_{\mathbb{C}} e^{isz} \frac{d\mu(x)}{k(z)} .$$

Now, if $s = \sigma + i\tau$ is such that $|\tau| \leq \frac{\varepsilon_1}{3}$ and $|\sigma|$ is bounded, the integral (27) still converges. This shows that every real analytic function can be analytically extended to the whole strip $|\tau| \leq \frac{\varepsilon_1}{3}$, which is obviously false.

Comparing Example 3 with Proposition 1 (cf. Remark 5) leads to the following unsolved

Problem. Is the division problem solvable in the space $\mathcal{R}$ of real analytic functions? *

---

*Added in the proofs: For n = 2 this has just been answered in the affirmative ny Ennio De Giorgi and Lamberto Cattabriga (cf. their forthcoming paper, "Una dimostrazione diretta dell' esistenza di soluzioni analitiche nel piano reale di equazioni a derivate parziali a coefficianti costanti").

CHAPTER II

Examples of AU-spaces

§1. THE BEURLING SPACES $\mathcal{D}_\omega$, $\mathcal{D}'_\omega$.

In this section we shall study an important class of function spaces $\mathcal{D}_\omega$ and their duals $\mathcal{D}'_\omega$ depending on a parameter ω taken from a certain family $\mathfrak{M}$ defined below. This class was first considered by A. Beurling [8]. The Schwartz spaces $\mathcal{D}$ and $\mathcal{D}'$ represent a special, and in a well-defined sense, extreme case of Beurling spaces (cf. Remark 1 below).[1]

<u>Definition 1</u>. $\mathfrak{M}$ denotes the class of all real valued functions ω, defined on the space $\mathbb{R}^n$, such that

$$(\alpha)\quad 0 = \omega(0) = \lim_{x\to 0}\omega(x) \leq \omega(\xi+\eta) \leq \omega(\xi) + \omega(\eta) \qquad (\forall \xi,\eta \in \mathbb{R}^n)\ ;$$

$$(\beta)\quad J_n(\omega) = \int_{\mathbb{R}^n} \frac{\omega(\xi)\,d\xi}{(1+|\xi|)^{n+1}} < \infty \quad ;$$

(γ) for some real number $\underline{a}$ and a positive number $\underline{b}$,

$$\omega(\xi) \geq a + b\log(1+|\xi|) \qquad (\forall \xi \in \mathbb{R}^n).$$

<u>Definition 2</u>. Given $\omega \in \mathfrak{M}$ and K any compact set in $\mathbb{R}^n$, let $\mathcal{D}_\omega(K)$ be the vector space of functions $\phi \in L^1(\mathbb{R}^n)$ with support in K and such that, for all $\lambda > 0$,

$$(\lambda) \qquad |\phi|_\lambda^{(\omega)} = |\phi|_\lambda = \int_{\mathbb{R}^n} |\hat{\phi}(\xi)| e^{\lambda\omega(\xi)} d\xi < \infty .$$

The space $\mathcal{D}_\omega(K)$ is equipped with the topology generated by the system of norms $\{|\cdot|_\lambda^{(\omega)}\}_{\lambda > 0}$. $\mathcal{D}_\omega(K)$ is obviously a Fréchet space, and by (γ), the elements of $\mathcal{D}_\omega(K)$ are C_0^∞-functions (cf. [9]). Let $\{K_s\}_{s \geq 1}$ be any sequence of compact sets exhausting $\mathbb{R}^n$. The space $\mathcal{D}_\omega = \mathcal{D}_\omega(\mathbb{R}^n)$ is then defined as the inductive limit

$$(1) \qquad \mathcal{D}_\omega = \lim_{s \to \infty} \operatorname{ind} \mathcal{D}_\omega(K_s) .$$

The definition of $\mathcal{D}_\omega$ is actually independent of the sequence $K_s \nearrow \mathbb{R}^n$. Therefore we shall always take for K_s the balls $K_s = \{x: |x| \leq R_s\}$ where $\{R_s\}$ is some fixed sequence such that $0 < R_s \nearrow +\infty$. The space $\mathcal{D}_\omega$ is called the <u>Beurling space of ω-test-functions</u>. Similarly, the dual $\mathcal{D}'_\omega$ of $\mathcal{D}_\omega$ is called the space of all <u>Beurling ω-distributions</u>. $\mathcal{T}_\omega(\text{ind})$ will denote the topology of $\mathcal{D}_\omega$ (cf. (1)).

<u>Remarks</u>: 1. Actually,* conditions (α), (β) and (γ) imposed on functions ω are very natural. Thus, condition (α) guarantees that $\mathcal{D}_\omega$ is an algebra under the pointwise multiplication and

$$(2) \qquad |\phi\psi|_\lambda \leq (2\pi)^{-n} |\phi|_\lambda \cdot |\psi|_\lambda$$

for all $\lambda > 0$ and ϕ,ψ in $\mathcal{D}_\omega$. Restriction (β) is obviously a Denjoy-Carleman type of condition, i.e. (β) is equivalent to the non-triviality of the space $\mathcal{D}_\omega$ (cf. [9] and Chap. III,

*for proofs, cf. [9].

§2 below). Condition (γ) is equivalent to the inclusion $\mathcal{D}_\omega \subset C_0^\infty(\mathbb{R}^n)$. Moreover, if we set $\omega_0(\xi) = \log(1+|\xi|)$, then it is easy to see that $\mathcal{D}_{\omega_0}$ is just the Schwartz space $\mathcal{S}$; and, for any $\omega \in \mathfrak{M}$, the space $\mathcal{D}_\omega$ is densely embedded into $\mathcal{S} = \mathcal{D}_{\omega_0}$. Therefore the Schwartz space $\mathcal{S}$ is the largest possible space of Beurling test functions. This also shows that the Fourier transform $\hat{\phi}$ of any Beurling test function is an entire function. Moreover, for each $\omega \in \mathfrak{M}$, $\mathcal{D}'_\omega \supset \mathcal{S}'$. There are some other function spaces which can also be obtained as $\mathcal{D}_\omega$ for some special choice of ω. Thus, for instance, by taking $\omega(\xi) = |\xi|^{1/\gamma}$, $\gamma > 1$, we obtain the Gevrey classes $\mathcal{E}_B \cap \mathcal{D}$ where B is the sequence $\{k^\gamma\}_{k \geq 1}$ and $\mathcal{E}_B$ is the space studied in Chapter III below.

2. If K is a compact set in $\mathbb{R}^n$, the supporting function H_K of K is defined as

(3)
$$H_K(\eta) = \max_{x \in K} \langle x,\eta\rangle \qquad (\eta \in \mathbb{R}^n).$$

It is shown in [9] that on each $\mathcal{D}_\omega(K_s)$ the system of norms $\{|\cdot|_\lambda^{(\omega)}\}_{\lambda>0}$ defined in (λ) is equivalent to either of the two systems $\{\|\cdot\|_\lambda^{(\omega)}\}_{\lambda>0}$ and $\{|||\cdot|||_{\lambda,s}^{(\omega)}\}_{\lambda,s>0}$, where the corresponding norms are defined as follows:

($\lambda\lambda$)
$$\|\phi\|_\lambda^{(\omega)} = \sup_{\xi \in \mathbb{R}^n} (|\hat{\phi}(\xi)|e^{\lambda\omega(\xi)}) ,$$

($\lambda\lambda\lambda$)
$$|||\phi|||_{\lambda,s}^{(\omega)} = \sup_{\zeta \in \mathbb{C}^n} [|\hat{\phi}(\zeta)|\exp(\lambda\omega(\xi) - H_{K_s}(\eta) - \tfrac{1}{4}|\eta|] .$$

where $\zeta = \xi + i\eta \in \mathbb{C}^n$. Since the spaces $\mathcal{D}_\omega$ are defined in terms of the Fourier transform, we shall often transfer different notions from $\mathcal{D}_\omega$ to $\hat{\mathcal{D}}_\omega$ without mentioning it explicitly. Thus, for instance, it is clear how to define the norms (λ)-$(\lambda\lambda\lambda)$ for f entire, $f \in \hat{\mathcal{D}}_\omega$. In this connection the following result will be useful: Let f be entire and such that $|||f|||^{(\omega)}_{\lambda,s} < \infty$ for some ω and $s = s_o$ fixed, and for all $\lambda > 0$. Then $f = \hat{\phi}$, where $\phi \in \mathcal{D}_\omega$ and $\operatorname{supp} \phi \subseteq \{x: |x| \leq R_s + \frac{1}{4}\}$. (The converse is trivial.) This is, of course, the Paley-Wiener theorem for the spaces $\mathcal{D}_\omega$ [9].

3. As can be easily seen, the Beurling spaces have the same geometric properties as the Schwartz spaces $\mathcal{D}$, $\mathcal{D}'$: For each $\omega \in \mathfrak{M}$, the spaces $\mathcal{D}_\omega$, $\mathcal{D}'_\omega$ are bornological Montel (and thus also barreled and reflexive) spaces, etc.

For later purposes it will be convenient to replace each ω by another function $\tilde{\omega}$ defining the same space $\mathcal{D}_\omega$:

Lemma 1. For each $\omega \in \mathfrak{M}$, set $\tilde{\omega} = 1 + \rho * \omega$ where ρ is a fixed C_o^∞-function with $\operatorname{supp} \rho = \{x: |x| \leq \varepsilon\}$ and $\int_{\mathbb{R}^n} \rho(\xi) d\xi = 1$; the positive number ε is taken so small that $\omega(\xi) \leq 1$ for $|\xi| \leq \varepsilon$. Then $\tilde{\omega}$ is a C_o^∞-function such that

$$(\alpha_1) \qquad 1 \leq \tilde{\omega}(\xi + \eta) \leq \tilde{\omega}(\xi) + \tilde{\omega}(\eta) ;$$

and, for any multiindex $\iota = (\iota_1, \ldots, \iota_n)$, there is a constant $T_\iota > 0$ such that

$$(\alpha_2) \qquad |D^\iota \tilde{\omega}(\xi)| \leq T_\iota (1 + |\xi|)^{n+1} .$$

Moreover, since for all ξ, $|\omega(\xi) - \tilde{\omega}(\xi)| \leq 2$, the spaces $\mathcal{D}_\omega$ and $\mathcal{D}_{\tilde{\omega}}$ coincide.

Proof. By the subadditivity of ω,

$$\begin{aligned}\tilde{\omega}(\xi + \eta) &= 1 + \int \omega(\xi + \eta - t)\rho(t)dt \\ &\leq 1 + \int \omega(\eta - t)\rho(t)dt + \int \omega(\xi)\rho(t)dt \\ &\leq \tilde{\omega}(\eta) + \int [\omega(\xi - t) + \omega(t)]\rho(t)dt \\ &\leq \tilde{\omega}(\eta) + \tilde{\omega}(\xi) .\end{aligned}$$

Property (α_2) follows from (β). Indeed,

$$\begin{aligned}|D^\iota\tilde{\omega}(\xi)| &\leq 1 + \int \frac{\omega(\xi-t)}{(1+|\xi-t|)^{n+1}} (1 + |\xi-t|)^{n+1}|D^\iota\rho(t)|dt \\ &\leq 1 + T'_\iota J_n(\omega)(1 + |\xi|)^{n+1}, \qquad \text{etc.}\end{aligned}$$

Remark 4. From now on, ω will always stand for its modification $\tilde{\omega}$ defined above.

The main objective of this section is to prove the following theorem.

Theorem 1. The space $\mathcal{D}'_\omega$ of Beurling ω-distributions is an AU-space.

The theorem will follow from Propositions 1, 2, and 3 below, which are interesting in their own right. We start by introducing some additional topologies on the space $\mathcal{D}_\omega$. The first one is defined in a somewhat geometric fashion:

<u>Topology</u> $\mathcal{T}_\omega(\mathcal{g})$. For any positive constants C,λ, and arbitrary sequences $r_j \nearrow \infty$ and $a_j \nearrow \infty$, $j=0,1,\ldots,a_o=0$, let

$$(4) \qquad \Lambda_j = \{\zeta \in \mathbb{C}^n : a_j\omega(\xi) \le |\eta| \le a_{j+1}\omega(\xi)\}$$

and

$$(5) \quad \mathcal{U}(C,\lambda,\{r_j\},\{a_j\}) = \{\phi \in \mathcal{D}_\omega : \sup_{\zeta \in \Lambda_j} [|\Phi(\zeta)\exp(\lambda\omega(\xi)-r_j|\eta|)] \le C \quad \text{for } j=0,1,\ldots\}$$

Each set $\mathcal{U}$ of this form is absolutely convex and absorbing. Indeed, let ϕ be any function in $\mathcal{D}_\omega$. Then by Remark 2, there exists a positive constant A such that, for any $\delta > 0$,

$$|\Phi(\zeta)| \le C_\delta \exp(-\delta\omega(\xi) + A|\eta|)$$

for some $C_\delta > 0$. For j_o large, $r_{j_o} \ge A$; hence for $\delta \ge \lambda$, the function $(CC_\delta^{-1})\phi$ satisfies the inequalities defining the set $\mathcal{U}(C,\lambda,\{r_j\},\{a_j\})$ for $j \ge j_o$. In the remaining strips, we have $|\eta| \le a_{j_o}\omega(\xi)$, since $\omega \ge 1$. Therefore, by choosing $\delta \ge \lambda + Aa_{j_o}$,

$$|\Phi(\xi)| \le C_\delta\exp(-\delta\omega(\xi) + A|\eta|) \le C_\delta\exp(-\lambda\omega(\xi))$$

for $\zeta \in \Lambda_1 \cup \ldots \cup \Lambda_{j_o-1}$. Hence $(CC_\delta^{-1})\phi \in \mathcal{U}(C,\lambda,\{r_j\},\{a_j\})$.

This shows that there exists an l.c. topology $\mathcal{T}_\omega(\mathcal{g})$ on $\mathcal{D}_\omega$

having for the basis of neighborhoods of the origin the system of all sets $\mathcal{U}$ of the form (5).

Topology $\mathcal{T}_\omega(\mathcal{K})$. Let $\{H_s\}_{s\geq 1}$ be any concave sequence of positive numbers, $H_s \nearrow \infty$, $H_s/s \to 0$. Fix a positive number μ and a bounded sequence $\{\varepsilon_s\}_{s\geq 1}$ of positive numbers. Then the series

$$(6) \qquad k(\zeta) = k(\{H_s\};\{\varepsilon_s\};\mu;\zeta) = \sum_{s=1}^{\infty} \varepsilon_s \exp[-(s+\mu)\omega(\xi) + H_s|\eta|]$$

is locally uniformly convergent in $\mathbb{C}^n$ and defines a majorant in the sense of Chap. I. $\mathcal{K} = \mathcal{K}(\omega)$ will denote the system of all such series k. For each $k \in \mathcal{K}$, the set $\mathcal{V}(k)$, defined by

$$(7) \qquad \mathcal{V}(k) = \{\phi \in \mathcal{D}_\omega : |\hat{\phi}(\zeta)| \leq k(\zeta) \quad (\forall \zeta \in \mathbb{C}^n)\}$$

is clearly absolutely convex and absorbing. Hence, all sets $\mathcal{V}(k)$, $k \in \mathcal{K}$, define on $\mathcal{D}_\omega$ an l.c. topology which will be denoted by $\mathcal{T}_\omega(\mathcal{K})$.

Topology $\mathcal{T}_\omega(\Sigma)$. This topology is determined by the basis of neighborhoods $\mathcal{W}(k)$, $k \in \mathcal{K}(\omega)$, defined as follows:

$$(8) \qquad \mathcal{W}(k) = \Big\{\phi \in \mathcal{D}_\omega : \text{there exists a positive integer } N = N(\phi) \text{ such that } \phi \text{ can be written as } \phi = \sum_{j=1}^{N} \phi_j,\ \phi_j \in \mathcal{D}_\omega, \text{ and for all } \zeta, j,\ |\hat{\phi}_j(\zeta)| \leq \varepsilon_j \exp[-(j+\mu)\omega(\xi) + H_j|\eta|]\Big\}.$$

Proposition 1. For each $\omega \in \mathfrak{M}$,

$$\mathcal{T}_\omega(\text{ind}) = \mathcal{T}_\omega(g) = \mathcal{T}_\omega(\mathcal{K}) = \mathcal{T}_\omega(\Sigma) .$$ [2]

The proof will be divided into three steps:

1. $\mathcal{T}_\omega(\text{ind}) = \mathcal{T}_\omega(\Sigma)$. Each set $\mathcal{W}(k)$ of the form (8) absorbs all bounded sets in the space $\mathcal{D}_\omega = (\mathcal{D}_\omega, \mathcal{T}_\omega(\text{ind}))$. Indeed, each bounded set M in the latter space is bounded in some $\mathcal{D}_\omega(K_s)$, i.e. for some positive constant C_λ and all $\lambda > 0$,

$$\sup\{|\hat{\phi}(\zeta)| : \phi \in M, \zeta \in \mathbb{C}^n\} \leq C_\lambda \exp[-\lambda\omega(\xi) + H_{K_s}(\eta) + \frac{|\eta|}{4}] .$$

If we take N so large that

$$H_{K_s}(\eta) + \frac{|\eta|}{4} \leq H_N|\eta| ,$$

and set $\lambda = N + \mu$, then $cM \subset \mathcal{W}(k)$ for some positive c. Since the space $\mathcal{D}_\omega$ is bornological (cf. Remark 3), this proves that the topology $\mathcal{T}_\omega(\Sigma)$ is coarser than $\mathcal{T}_\omega(\text{ind})$. To prove the opposite relation, let $\mathcal{Z}$ be a convex neighborhood of the origin in the topology $\mathcal{T}_\omega(\text{ind})$. Then, for some $\delta_s \searrow 0$ and positive integers $\lambda_s' \nearrow \infty$,

$$\mathcal{Z} \cap \mathcal{D}_\omega(K_s) \supseteq \{\phi : \|\phi\|_{\lambda_s'} \leq \delta_s\} .$$

Let us define a new sequence of integers $\lambda_s \geq \lambda_s'$ as follows. First, we set $\lambda_1 = \lambda_1'$, $\lambda_2 = \lambda_2'$ and denote by p_1 the segment in the plane (λ, R) with endpoints $(0,0)$ and (λ_2, R_1) (R_s are the numbers of Def. 1). Let p_2^* be a halfray originating at the point (λ_2, R_1) and with slope being half the slope of p_1. Let A be the point on p_2^* for which $A = (\nu, R_2)$; λ_3 the integral part of $1 + \max\{\lambda_3', \nu\}$; and, p_2 the segment with endpoints (λ_2, R_1) and (λ_3, R_2). Continuing in the same way we obtain the broken line $p_1 \cup p_2 \cup p_3 \cup \ldots$, whose equation in the (t,R)-plane is $R = \psi(t)$. Obviously $\lambda_s \geq s$, and ψ is a concave function such that $\psi(t) \nearrow \infty$, $\psi(t)/t \to 0$. Furthermore, let $\mu = \lambda$, $H_s = \psi(s)$ and $\varepsilon_s = \delta_s 2^{-s}$. We claim that for $k = k(\{H_s\};\{\varepsilon_s\};\mu)$ (cf. (6)), $\mathcal{W}(k) \subset \mathfrak{X}$.

Let ϕ be any element in $\mathcal{W}(k)$. Then, for some integer $N = N(\phi)$, ϕ can be decomposed into the sum

$$\phi = \sum_{j=1}^{N} \phi_j = \sum_{j=1}^{N} \frac{1}{2^j}(2^j\phi) ,$$

where for each $\zeta \in \mathbb{C}^n$ and $j=1,2,\ldots,N$,

$$|\hat{\phi}_j(\zeta)| \leq \varepsilon_j \exp[H_j|\eta| - (j+\mu)\omega(\xi)] .$$

By the above construction, $H_j < R_s$ when $\lambda_s \leq j < \lambda_{s+1}$. Therefore by the Paley-Wiener theorem (cf. Remark 2), supp $\phi_j \subset K_s$. Moreover, $\|\phi_j\|_{\lambda_s'} \leq 2^{-j}\delta_j$. For the remaining indices j, $1 \leq j < \lambda_1$,

$$|\hat{\phi}_j(\xi)| \leq \delta_j 2^{-j} e^{-\mu\omega(\xi)} \leq 2^{-j}\delta_1 e^{-\lambda_1\omega(\xi)} .$$

This shows that, for all $j=1,2,\ldots,N$, $2^j\phi_j \in \mathfrak{X}$. Convexity of $\mathfrak{X}$ yields $\phi \in \mathfrak{X}$, and the equality $\mathcal{T}_\omega(\text{ind}) = \mathcal{T}_\omega(\Sigma)$ follows.

2. $\mathcal{T}_\omega(\mathcal{G}) = \mathcal{T}_\omega(\mathcal{K})$. In order to show that $\mathcal{T}_\omega(\mathcal{K})$ is coarser than $\mathcal{T}_\omega(g)$, one has to find, for each $k \in \mathcal{K}$, a set $\mathcal{U}(C,\lambda,\{r_j\},\{a_j\})$ contained in $\mathcal{V}(k)$. We claim that it suffices to choose $C = \min(1,\varepsilon_1)$, $\lambda = \mu+1$, $r_o = \frac{1}{2}H_1$, $r_s = H_s$ for $s \geq 1$ and a sequence $a_s \nearrow \infty$ such that

$$a_s \geq \frac{s - \log \varepsilon_{s+1}}{H_{s+1} - H_s} . \tag{9}$$

Indeed, given any $\phi \in \mathcal{U}(\ldots)$, we shall show that

$|\Phi(\zeta)| \leq k(\zeta)$ for all $\zeta \in \mathbb{C}^n$. If $\zeta \in \Lambda_o$, then

$$|\Phi(\zeta)| \leq \varepsilon_1 \exp[-\lambda\omega(\xi) + |\eta| \frac{H_1}{2}] < k(\zeta) .$$

If $\zeta \in \Lambda_j$, $j \geq 1$, then by (9), the function $|\Phi(\zeta)|$ is bounded by the term in the series k, for which $s=j+1$.

To prove the converse we begin by constructing an auxiliary function $p(t)$, $t \in \mathbb{R}$, which will be a differentiable, convex and even function on $\mathbb{R}$. Set $r_o = s_o = \alpha_o = p(0) = 0$, $s_1 = 1$, and construct the function p first on the interval $[-r_1,r_1]$ so that, in addition to the above properties on this interval, p will also satisfy the following conditions:

(i) $p(r_1) = s_1$;

(ii) if $(\alpha_1,-1)$ is the normal vector to the graph of the function p at the point (r_1,s_1), then $\alpha_1 > a_1$ and for some integer q_1, $\alpha_1 = a_{q_1}$. In particular $q_1 > 1$. If the integers $s_1 < s_2 < \ldots < s_m$ and the function $p(t)$ on $[-r_m,r_m]$ have already been defined so that

$$(10) \qquad p(r_j) = s_j , \qquad (j=1,\ldots,m)$$

and if $(\alpha_j,-1)$ is the normal vector to the graph of p at the point (r_j,s_j), then for some q_j,

$$(11) \qquad \alpha_j = a_{q_j} > a_j .$$

It is clear that the construction can be continued to finally yield function p such that

$$(12) \quad B_j \stackrel{\text{def}}{=} \{x : |x_j| \leq r_j\} = \{x : p(|x|) \leq s_j\} ,$$

and

(13) $$\alpha_j \nearrow \infty.$$

If the sequence $\{H_s\}$ is defined by the conditions

(14) $$p(H_j) = j \qquad (j \geq 1) ,$$

then by (13), the sequence $\{H_j\}_{j\geq 1}$ is concave, $H_j \nearrow \infty$, $H_j/j \to \infty$, and by (10), $H_{s_\ell} = r_\ell$. Set

(15) $$\mu = \lambda + a_1$$

and choose ε_s positive so that

(16) $$\sum_{s=1}^{\infty} \varepsilon_s < \frac{C}{2} .$$

For any ζ in $\mathbb{C}^n$ we write $\zeta = \xi + i\eta = \xi + i\Theta\omega(\xi)$. Then, for some integers ℓ and q,

(17) $$\alpha_\ell \leq |\Theta| < \alpha_{\ell+1} \text{ and } a_q \leq |\Theta| < a_{q+1}, \text{ i.e. } \zeta \in \Lambda_q .$$

Suppose first $\Theta \neq 0$. We claim that

(18) $$k(\zeta) \leq C\exp(-\lambda\omega(\xi) + r_{\ell+1}|\eta|) .$$

Indeed, if k is written as

$$k = \sum_{s=1}^{s_\ell} + \sum_{s>s_\ell} ,$$

then

$$(19) \qquad \sum_{s=1}^{s_\ell} \ldots \leq \frac{C}{2} \exp(-\lambda\omega(\xi) + r_\ell|\eta|) .$$

For the estimate of the second sum, the geometric properties of the function $p(|x|)$ have to be used. First, as this function is symmetric, relations (12) and (17) imply that there exists a point x such that, $r_\ell \leq |x| < r_{\ell+1}$ and $(\theta,-1)$ is the normal vector to the graph of the function $p(|x|)$ at the point $(x,p(|x|))$. Moreover the vectors x and θ are collinear, i.e. $x = |x|\theta/|\theta|$. The convexity of p then implies the inequality

$$(20) \qquad [(y,p(|y|)) - (x,p(|x|))] \cdot (\theta,-1) \leq 0$$

for any $y \in \mathbb{R}^n$. In particular, for $y = H_s\theta/|\theta|$ $(s > s_\ell)$, we get $p(|y|) = s$ and

$$(21) \qquad (H_s - |x|)|\theta| \leq s - p(|x|) \leq s - s_\ell ,$$

whence

$$(22) \quad \begin{cases} \displaystyle\sum_{s=s_\ell+1}^{\infty} \varepsilon_s \exp\{(\lambda-s-\mu)\omega(\xi) + (H_s-r_{\ell+1})|\eta|\} \\ \leq \sum \varepsilon_s \exp\{-s\omega(\xi) + (H_s-|x|)|\theta|\omega(\xi)\} \qquad \text{(by (21):)} \\ \leq \left(\sum \varepsilon_s\right) e^{-s_\ell\omega(\xi)} \leq \frac{C}{2} . \end{cases}$$

Inequality (18) now follows from (19) and (22). The assumption $\Theta \neq 0$ is automatically satisfied if $q \geq 1$. In this case we claim that

$$(23) \qquad k(\zeta) \leq C\exp(-\lambda\omega(\xi) + r_q|\eta|) \qquad (\zeta \in \Lambda_q;\ q \geq 1).$$

Here we have to distinguish two cases: $\ell = 0$ and $\ell \geq 1$. For $\ell = 0$, $r_{\ell+1} = r_1 \leq r_q$ and (23) follows from (18). If $\ell \geq 1$, then by (11), $\alpha_\ell = a_{q_\ell}$. However, by the construction, $q_1 > 1$; hence $q_\ell > \ell$. Comparison of bounds in (17) gives $a_{q_\ell} \leq a_q$. Thus $\ell+1 \leq q_\ell \leq q$, and (23) follows again from (18). The only region which still has to be checked is Λ_0. There, however, $|\eta| \leq a_1\omega(\xi)$, and (16) and (17) give

$$(24) \qquad k(\zeta) \leq Ce^{-\lambda\omega(\xi)} \qquad (\zeta \in \Lambda_0).$$

Inequalities (23) and (24) prove the inclusion $\mathcal{V}(k) \subset \mathcal{U}(C,\lambda,\{r_s\},\{a_s\})$, which completes the proof of part 2.

3. $\mathcal{T}_\omega(\text{ind}) = \mathcal{T}_\omega(g)$. Since $\mathcal{T}_\omega(\text{ind})$ is barreled (cf. Remark 3), the topology $\mathcal{T}_\omega(g)$ is clearly coarser than $\mathcal{T}_\omega(\text{ind})$.

Now let $\mathfrak{X}$ be a closed convex neighborhood in the topology $\mathcal{T}_\omega(\text{ind})$. Then, for some $\varepsilon_s \searrow 0$ and $\nu_s \nearrow \infty$,

$$\mathfrak{X} \cap \mathcal{D}_\omega(B(s)) \supset \left\{\phi \in \mathcal{D}_\omega(B(s)) : \sup_{\xi \in R^n} \left(|\hat{\phi}(\xi)| e^{\nu_s\omega(\xi)}\right) \leq \varepsilon_s\right\},$$

where we denoted $B(s) = \{x : |x|_1 = |x_1|+\ldots+|x_n| \leq s\}$. We claim that for a convenient choice of the parameters, $\mathcal{U}(C,\lambda,\{r_j\},\{a_j\}) \subset \mathfrak{X}$. Let $\mathcal{M}$ be the set of all lattice points $M = (m_1,\ldots,m_n)$ in R^n such that $|M| \geq n\sqrt{n}$. Furthermore, let us set

$$S_M = \left\{x : |x - M| < \frac{5\sqrt{n}}{8}\right\}$$

for $M \in \mathcal{M}$, and

$$S_0 = \{x : |x| < (n+1)\sqrt{n}\}.$$

By [9] there exists a partition of unity $\{\alpha_M\}$, $\alpha_M \in \mathcal{D}_\omega$, subordinate to the covering $\{S_M : M \in \mathcal{M} \cup \{0\}\}$ of $\mathbb{R}^n$. Then, for any positive σ, we have by Remark 2,

(25) $$|\hat{\alpha}_M(\zeta)| \le C_{\sigma,M} \exp(-\sigma\omega(\xi) + H_{S_M}(\eta) + \frac{1}{4}|\eta|) .$$

As above, we have set

$$\phi = \sum \frac{1}{2^{|M|_1}} \left(2^{|M|_1} \alpha_M\phi\right)$$

for each $\phi \in \mathcal{U}(\ldots)$. The parameters defining the neighborhood $\mathcal{U}(\ldots)$ will be chosen so that, for every M,

(26) $$2^{|M|_1+n} \alpha_M\phi \in \mathfrak{X} .$$

(Let us recall that $\Sigma 2^{-|M|_1} = 2^n$). Then convexity of $\mathfrak{X}$ will imply $\phi \in \mathfrak{X}$.

Obviously, for $M \in \mathcal{M}$, $\operatorname{supp}(\alpha_M\phi) \subset B(2|M|_1)$; and, $\operatorname{supp}(\alpha_0\phi) \subset B(n^2+n)$. Since $\phi \in \mathcal{U}(\ldots)$, we obtain

(27) $$\begin{cases} |\widehat{\phi\alpha}_0(\xi)| \le \int_{\mathbb{R}^n} |\hat{\phi}(\xi-t)||\hat{\alpha}_0(t)|dt \\ \le CC_{\sigma,0}\int \exp[-\lambda\omega(\xi-t) - \sigma\omega(t)]dt \\ \le CC_{\sigma,0}e^{-\sigma\omega(\xi)} \int \exp[(\sigma-\lambda)\omega(\xi-t)]dt , \end{cases}$$

where we used the subadditivity of ω.

Now let us set $\sigma = \nu_{n^2+n}$ and

(28) $$\lambda = \nu_{n^2+n} + \frac{n+1}{b}$$

where b is the constant from condition (γ), Def. 1. We also fix $C > 0$ so that

$$(29) \qquad 2^n C C_{\nu_{n^2+n},0} \int_{\mathbb{R}^n} \frac{dt}{(1+|t|)^{n+1}} \leq \varepsilon_{n^2+n} .$$

Then inclusion (26) is verified for $M = 0$.

In order to estimate those terms in (26) for which $|M|_1 = s > 0$, we have to shift the integration in the convolution $\widehat{\alpha_M \phi}$ from $\mathbb{R}^n$ to the variety $\Gamma_\xi = \{\tau = t+i\eta \in \mathbb{C}^n : \eta = -\frac{M}{|M|} a_s\omega(\xi-t),\ t \in \mathbb{R}^n\}$. By the Cauchy-Poincaré formula, this can be done provided

$$(30) \qquad \lim_{|t| \to \infty} |t|^{n-1} \int_0^{a_s\omega(\xi-t)} |\hat{\phi}(\xi-t+i\tfrac{M}{|M|}v)\, \hat{\alpha}_M(t-i\tfrac{M}{|M|}v)|\,dv = 0 .$$

However, relation (30) follows from obvious estimates:

$$|t|^{n-1} \int_0^{a_s\omega(\xi-t)} \cdots \leq C_{\phi,\rho}|t|^{n-1} a_s\omega(\xi-t)\exp[-\rho\omega(\xi-t) + Ra_s\omega(\xi-t)] \times$$

$$\times\ C_{\sigma,M}\exp[-\sigma\omega(t) + (s+1)a_s\omega(\xi-t)] \to 0 .$$

Here we used the inclusion $\operatorname{supp} \phi \subset \{x : |x| \leq R - \frac{1}{4}\}$, and the notation $\rho = Ra_s + (s+1)a_s + 1$, $\sigma = \frac{n}{b}$. Therefore,

$$(31) \qquad \widehat{\alpha_M \phi}(\xi) = \int_{\Gamma_\xi} \hat{\phi}(\xi-\tau)\hat{\alpha}_M(\tau)\,d\tau .$$

In the last integral, $|d\tau| \leq (1 + |\xi-t|)^{n^2+n}(1 + Ta_s)^n$, where T is the corresponding constant from Lemma 1. Therefore, using (25) and the fact that $\xi-\tau \in \Lambda_s$ for $\tau \in \Gamma_\xi$, we get from (31) (cf. (1))

$$(32)\quad \begin{cases} |\widehat{\alpha_M\phi}(\xi)| \leq CC_{\sigma,M}(1+Ta_s)^n e^{-\sigma\omega(\xi)} \times \\ \times \int_{\mathbb{R}^n} (1+|\xi-t|)^{n^2+n} \exp\left\{\omega(\xi-t)\left[\sigma\cdot\lambda+a_s\left(\frac{1}{4}+r_s+H_{S_M}\left(-\frac{M}{|M|}\right)\right)\right]\right\} dt \end{cases}$$

Let $\sigma = \nu_{2s}$. The sequences $\{r_j\}$ and $\{a_j\}$ still remain to be chosen. Clearly,

$$H_{S_M}(-M/|M|) = -|M| + \frac{5}{8}\sqrt{n} \quad \text{and} \quad n\sqrt{n} \leq |M| \leq |M|_1 = s \leq n|M| .$$

Thus, if

$$r_s \overset{\text{def}}{=} \frac{s}{n} - \frac{15}{16}\sqrt{n} ,$$

then $0 < r_s \nearrow \infty$, and

$$(33)\qquad r_s \leq |M| - \frac{15}{16}\sqrt{n} .$$

From here we obtain (cf. (32))

$$(34)\qquad \frac{1}{4} + r_s + H_{S_M}(-M/|M|) \leq -\frac{1}{16} .$$

Let Q be the constant

$$Q = Ce^{-(n+1)^2 a} \max_{\{M:|M|_1 = s\}} C_{\sigma,M} \int_{\mathbb{R}^n} \frac{dt}{(1+|t|)^{n+1}} ,$$

where $\underline{a}$ is the constant from (γ), Def. 1. If a_s is chosen so large that

(35) $$a_s > 32(\sigma - \lambda + (n+1)^2 b^{-1}) ,$$

then, by (34) and (35), inequality (32) yields

(36) $$|\alpha_M \hat{\phi}(\xi)| \leq Q(1 + Ta_s)^n \exp\left[- \frac{a_s}{32} - \nu_{2s}\omega(\xi)\right] .$$

By taking a_s even larger we can achieve that

(37) $$Q(1 + Ta_s)^n \exp(- \frac{a_s}{32}) \leq \varepsilon_{2s} 2^{-s-n} ,$$

and (36) then implies (26).

So far all parameters defining the neighborhood $\mathcal{U}(C,\lambda,\{r_j\},\{a_j\})$ have already been chosen except for the first few values of a_j and r_j. However these can be defined arbitrarily as long as the sequences $\{a_j\},\{r_j\}$ will remain positive and strictly increasing. The proof of part 3 is complete and Proposition 1 follows.

<u>Corollary</u>. Given any function k in the family $\mathcal{K}=\mathcal{K}(\omega)$,

$$k(\zeta) = \sum_{s=1}^{\infty} \varepsilon_s \exp[-(s+\mu)\omega(\xi) + H_s|\eta|] ,$$

there exists another majorant $\tilde{k} \in \mathcal{K}(\omega)$,

$$\tilde{k}(\zeta) = \sum_{s=1}^{\infty} \tilde{\varepsilon}_s \exp[-(s+\mu)\omega(\xi) + H_s|\eta|]$$

such that, if $\phi \in \mathcal{D}_\omega$ and $|\hat{\phi}(\zeta)| \leq \tilde{k}(\zeta)$ for all $\zeta \in \mathbb{C}^n$, then there are functions ϕ_j $(j=1,\ldots,N;\ N = N(\phi))$ in $\mathcal{D}_\omega$ such that

$\phi = \sum_{j=1}^{N} \phi_j$ and

$$|\hat{\phi}_j(\zeta)| \leq \varepsilon_j \exp[-(j+\mu)\omega(\xi) + H_j|\eta|] \quad (j=1,\ldots,N;\ \zeta \in \mathbb{C}^n) .$$

This statement is similar to a lemma due to A. Macintyre (cf. [11], p. 80). An interesting problem would be to find any estimate for $\tilde{H}_s, \tilde{\varepsilon}_s, \tilde{\mu}$ in terms of H_s, ε_s, μ. This would probably follow from a constructive way of proving that the topology $\mathcal{T}_\omega(\Sigma)$ is coarser than $\mathcal{T}_\omega(\mathcal{K})$. A similar problem was studied by B. A. Taylor in [49] who who used the technique of L^2-estimates of the $\bar{\partial}$-operator (cf. [29]). In our case, this does not seem to work.

Next we want to show that the family $\mathcal{K}(\omega)$ determines completely which C_0^∞-functions are elements of $\mathcal{D}_\omega$ (cf. property (A) in Chapter I).

Proposition 2. For each ω and $\mathcal{K} = \mathcal{K}(\omega)$ as above, $\mathcal{A}(\mathcal{K}(\omega)) = \hat{\mathcal{D}}_\omega$.

The proof of this statement depends on a lemma (see Lemma 2 below) which will be useful on several occasions in this section. We shall employ the following notation: $\mathcal{C}$ (or $\mathcal{C}_+$) denotes the class of all functions h which are concave, increasing to $+\infty$, continuously differentiable on $[0,\infty)$ and such that $h(0) \geq 0$ (or $h(0) > 0$ resp.) and $0 < h'(s) \leq \frac{1}{2s+1}$ for all $s \geq 0$.

Lemma 2. Let h be a function in $\mathcal{C}$ and p its inverse. Then, for all $a > 0$ and $b \geq 1$,

$$(38) \qquad \left.\begin{matrix} \displaystyle\sum_{s=0}^{\infty} e^{ah(s)-bs} \\[2ex] \displaystyle\sum_{s=0}^{\infty} e^{as-bp(s)} \end{matrix}\right\} \leq (5 + 2a)e^{ah(a)}$$

In the second inequality we assume $h(0) = 0$.

Proof. First, let us show that

$$\sum_{s=0}^{\infty} e^{ah(s)-bs} \leq (5 + \tfrac{3}{2}a)e^{ah(a)} . \tag{39}$$

In the proof of (39) a simple version of the Euler-Maclaurin formula will be needed:

Let α, β be integers, $\alpha < \beta$, $f(x)$ a continuously differentiable function in $[\alpha,\beta]$ and $\Phi(x)$ the function of period 1 such that $\Phi(x) = x - \frac{1}{2}$ for $0 < x < 1$. Then

$$\tfrac{1}{2}f(\alpha) + [f(\alpha+1)+\ldots+f(\beta-1)] + \tfrac{1}{2}f(\beta) = \int_{\alpha}^{\beta} f(x)dx + \int_{\alpha}^{\beta} f'(x)\Phi(x)dx . \tag{40}$$

From here follows

$$\begin{aligned}\sum_{s=0}^{N} e^{ah(s)-bs} &= \tfrac{1}{2}e^{ah(0)} + \tfrac{1}{2}e^{ah(N)-N} + \int_{0}^{N} e^{ah(s)-bs}ds \\ &\quad + \int_{0}^{N} (ah'(s)-b)e^{ah(s)-bs}\Phi(s)ds .\end{aligned} \tag{41}$$

Setting $bs = t$ in the last integral we obtain

$$\left\{\begin{aligned}\left|\int_{0}^{N}(\ldots)\right| &\leq \frac{1}{2b}\int_{0}^{Nb} [ah'(\tfrac{t}{b}) + b]\exp[ah(\tfrac{t}{b}) - t]dt \\ &\leq \frac{1}{2b}\int_{0}^{\infty}\left(\frac{a}{2(\frac{t}{b}) + 1} + b\right) e^{ah(t)-t}dt \\ &\leq \frac{a+1}{2}\int_{0}^{\infty} e^{ah(t)-t}dt .\end{aligned}\right. \tag{42}$$

(Here we used the obvious inequality $h(\frac{t}{b}) \leq h(t)$.) By (42), equation (41) implies

$$\sum_{s=0}^{\infty} e^{ah(s)-bs} \leq \frac{1}{2} e^{ah(0)} + \frac{a+3}{2} \int_0^{\infty} e^{ah(t)-t} dt . \tag{43}$$

Now it remains to estimate the last integral which can be written as $\int_0^a (\ldots) + \int_a^{\infty} (\ldots)$. However,

$$\int_0^a (\ldots) \leq e^{ah(a)} \int_0^a e^{-s} \leq e^{ah(a)} .$$

Moreover,

$$\int_a^{\infty} (\ldots) = [e^{ah(s)-s}]_{\infty}^{a} + \int_a^{\infty} ah'(s) e^{ah(s)-s} ds$$

$$\leq e^{ah(a)-a} + \frac{a}{2a+1} \int_a^{\infty} (\ldots) .$$

Hence,

$$\int_0^{\infty} e^{ah(s)-s} ds \leq 3e^{ah(a)} \tag{44}$$

and (39) follows from (43) and (44).

The proof of the second inequality in (38) is similar:

$$\sum_{s=0}^{N} e^{as-bp(s)} = \frac{1}{2}\left(1 + e^{aN-bp(N)}\right) + \int_0^N e^{as-bp(s)} ds$$

$$+ \int_0^N [a - bp'(s)] e^{as-bp(s)} \Phi(s) ds . \tag{45}$$

Furthermore, substituting $\sigma = p(s)$ in both integrals, we obtain

$$\sum_{s=0}^{\infty} e^{as-bp(s)} \leq \frac{1}{2} + \frac{a+3}{2} \int_0^{\infty} e^{ah(t)-t} dt . \tag{46}$$

Now it is clear that the second estimate in (38) follows along the same lines as the first one.

In order to prove Proposition 2, let f be a fixed entire function such that $f = \mathcal{O}(k)$ for each $k \in \mathcal{K}$. In particular, if h is any element in $\mathcal{C}_+$, the sequence $H_s = h(s)$ $(s=1,2,\ldots)$ combined with an arbitrary $\mu > 0$ and any bounded sequence $\{\varepsilon_s\}$ defines a majorant $k = k(\{H_s\};\{\varepsilon_s\};\mu) \in \mathcal{K}(\omega)$. Thus by (38),

$$|f(\zeta)|e^{\mu\omega(\xi)} \leq Ck(\zeta)e^{\mu\omega(\xi)} \leq C_1 \sum_{s=1}^{\infty} e^{|\eta|H_s - s\omega(\xi)}$$

$$\leq C_1\left(5 + \frac{3}{2}|\eta|\right) e^{|\eta|h(|\eta|)}$$

$$\leq C_2 e^{2|\eta|h(|\eta|)} ,$$

for some constants C, C_1, C_2 depending only on k and f. If we set $g(|\eta|) = \sup\{|\eta|^{-1}\log[C_2^{-1}(\mu\omega(\xi) + |f(\zeta)|)] : V(\eta/|\eta|);\ V\xi;\ V\mu\}$, then the last inequality can be written as

$$\sup_{|\eta|} \frac{g(|\eta|)}{h(|\eta|)} \leq 2 . \tag{47}$$

Since h was an arbitrary element in $\mathcal{C}_+$, we conclude from (47) (cf. [17]) that $g \leq B$ for some $B > 0$, i.e. for all $\zeta \in \mathbb{C}^n$ and any $\mu > 0$,

$$|f(\zeta)| \leq C_\mu e^{B|\eta| - \mu\omega(\xi)} \tag{48}$$

with $C_\mu = C_2$. The Paley-Wiener theorem (cf. Remark 2) then implies $f = \hat{\phi}$ for some $\phi \in \mathcal{D}_\omega$; and, this completes the proof of Proposition 2.

To complete the proof of Theorem 1, it remains to exhibit a suitable BAU-structure for $\mathcal{D}'_\omega$. Let $\mathfrak{S}$ be the class of all sequences $\{C_j\}_{j\geq 1}$ of positive numbers such that $1 = \omega(0) < C_{j+1} - C_j \nearrow \infty$. Given any $\{C_j\} \in \mathfrak{S}$ and arbitrary $A > 0$, $C > 0$, set

$\phi(\xi) = \inf_n (C_n - n\omega(\xi))$ and

$$m(\zeta) = m(\{C_j\};A;C;\zeta) = Ce^{\phi(|\xi|)+A|\eta|} . \tag{49}$$

Let $\mathcal{M}(\mathfrak{S})$ be the family of all such functions m. Furthermore, let $\mathcal{L}$ ($\mathcal{L}_c$ resp.) be the class of all positive functions $\Lambda(t)$ $(t \geq 1)$ for which $\lambda(t) \overset{\text{def}}{=} \Lambda(t)/t \to \infty$ when $t \to \infty$ (and λ concave resp.). Given any $\Lambda \in \mathcal{L}$ ($\Lambda \in \mathcal{L}_c$ resp.) and arbitrary $A > 0$, $C > 0$, let

$$m(\zeta) = m(\Lambda;A;C;\zeta) = Ce^{-\Lambda(\omega(\xi))+A|\eta|} . \tag{50}$$

Denote $\mathcal{M}(\mathcal{L})$ ($\mathcal{M}(\mathcal{L}_c)$ resp.) the family of all such functions m.

<u>Proposition 3</u>. Each of the three families $\mathcal{M}(\mathfrak{S})$, $\mathcal{M}(\mathcal{L})$, $\mathcal{M}(\mathcal{L}_c)$ is a BAU-structure for the space $\mathcal{D}'_\omega$ and satisfies condition (vi) of Def. 1, I.[3]

<u>Proof</u>. Let us first observe that each $\phi \in \mathcal{M}(\mathfrak{S})$ is continuous. Indeed, define $V_j = \{\xi : C_{j+1} \geq \omega(\xi) + C_j\}$, $V_0 = \emptyset$. The sequence $\{V_j\}_{j\geq 1}$ exhausts $\mathbb{R}^n$; and, for $\xi \in V_j \setminus V_{j-1}$, $\phi(\xi) = C_j - j\omega(\xi)$. The continuity of the function ϕ then follows. Let k be any series as in (6), and m an arbitrary element of $\mathcal{M}(\mathfrak{S})$. Then, for some s_1, $H_s \geq \Lambda$ for all $s \geq s_1$; and,

$$\sum_{s\geq s_1} \varepsilon_s \exp[-(s+\mu)\omega(\xi) - \phi(\xi)] = \sum_{s\geq s_1} \varepsilon_s \exp[-(s+\mu)\omega(\xi) + \max_n (n\omega(\xi) - C_n)]$$

$$\geq \sum_{s\geq s_1} \varepsilon_s \exp(-C_{[\mu]+1+s}) .$$

Hence, denoting the last term by $C/\tilde{C}$, we obtain $m(\zeta) \leq \tilde{C}k(\zeta)$ for all ζ. Conversely, let B be a bounded set in $\mathcal{D}_\omega$. Then by (1), B is a bounded subset of some $\mathcal{D}_\omega(K_{s_o})$, where $K_{s_o} = \{x : |x| \leq R_{s_o}\}$, i.e. $\sup_{f \in B} \|f\|_n^{(\omega)} \leq \tilde{C}_n$ (n=1,2,...). Let us choose $A = R_{s_o} + 1$, $C_n > \log \tilde{C}_n$, and $m = m(\{C_n\};A;...)$. Then $B \subset A(m;\alpha)$ for some $\alpha > 0$; condition (vi) is easy to check.

Now let $m \in \mathcal{M}(\mathcal{L})$ and k as above. We claim that $k(\zeta)/m(\zeta) \geq$ const. > 0. In fact, by (6) and (50),

$$(51) \qquad \frac{k(\zeta)}{m(\zeta)} = C^{-1} \sum_{s=1}^{\infty} \varepsilon_s \exp[(H_s - A)|\eta| - (s+\mu)\omega(\xi) + \omega(\xi)\lambda(\omega(\xi))] .$$

Thus, choosing s_o so that $H_{s_o} > A$, and $\Xi > 0$ so large that $\lambda(\omega(\xi)) \geq s_o + \mu$, for $|\xi| \geq \Xi$, we obtain from (51),

$$\frac{k(\zeta)}{m(\zeta)} \geq C^{-1}\varepsilon_{s_o} \min\left\{e^{-(s_o+\mu)\omega(\xi)} : |\xi| \leq \Xi\right\} > 0 .$$

To prove that the family $\mathcal{M}(\mathcal{L})$ is a BAU-structure, it suffices to find for each $m \in \mathcal{M}(\mathfrak{S})$ a function $m^* \in \mathcal{M}(\mathcal{L})$ such that

$$(52) \qquad m(\zeta) \leq \text{const. } m^*(\zeta) .$$

If m is given by (49) and $\{V_j\}$ as above, let us fix an arbitrary sequence of points $\xi_j, \xi_j \in \partial V_j$ for $j \geq 1$, and $\xi_o = 0$. Then $|\xi_j| \nearrow \infty$ and $\omega(\xi_j) = C_{j+1} - C_j$. Hence $C_n = C_1 + \omega(\xi_1) + \ldots + \omega(\xi_{n-1})$, and by the definition of ϕ,

$$(53)\quad \frac{\phi(\xi)}{\omega(\xi)} = \frac{C_n}{\omega(\xi)} - n \leq \frac{C_1 + \omega(\xi_1) + \ldots + \omega(\xi_{n-2})}{\omega(\xi_{n-1})} - (n-1) \quad (\xi \in V_n \setminus V_{n-1}).$$

Now, let $\mathfrak{S}'$ be any subclass of $\mathfrak{S}$ such that, for each $\{\tilde{C}_j\} \in \mathfrak{S}$, there is a sequence $\{C_j\} \in \mathfrak{S}'$, for which $C_j \geq \tilde{C}_j$. Since we already know that $\mathcal{M}(\mathfrak{S})$ is a BAU-structure, this will imply that the family $\mathcal{M}(\mathfrak{S}')$ is also a BAU-structure. Therefore, it suffices to prove (52) for all m from any family $\mathcal{M}(\mathfrak{S}')$ with the foregoing properties. In particular, let us choose as $\mathfrak{S}'$ the class of all sequences $\{C_j\}$ in $\mathfrak{S}$, which grow so rapidly that the last term in (53) is $\leq 2-n$ for all $n \geq 2$. Let $m = m(\{C_j\};\ldots)$ be a fixed function in $\mathcal{M}(\mathfrak{S}')$. Then for all ξ such that $\omega(\xi) = t$, $\phi(\xi)$ assumes the same value which we shall denote by $-\Lambda^*(t)$. Let $\Lambda(t)$ be defined for $t \geq \omega(0)$ as $\Lambda(t) = \Lambda^*(t) = c$, where the constant $\underline{c}$ was chosen so large that $\Lambda(t) > 0$; and, $\Lambda(t) \overset{\text{def}}{=} \Lambda(\omega(0))$ for $t \in [0,\omega(0))$. Then, by (53), $\phi(\xi)/\omega(\xi) \leq 2-n$, which implies $\Lambda \in \mathcal{L}$. Now the inequality (53) follows for $m^*(\zeta) = C^*\exp[-\Lambda(\omega(\xi)) + A|\eta|]$.

Finally, to show that $\mathcal{M}(\mathcal{L}_c)$ is also a BAU-structure for $\mathcal{D}'_\omega$, it suffices to show (cf. the discussion following (53)) that, for each $\lambda(t) = \Lambda(t)/t$, $\Lambda \in \mathcal{L}$, there exists a positive concave function λ^*, such that $\lambda^* \leq \lambda$ and $\lim_{t \to \infty} \lambda^*(t) = \infty$. However, this is easy to see (cf. [17], Lemma 6). Furthermore, the verification of the rest of condition (v) of Def. 1,I is straightforward. Thus Proposition 3 is proved; and, this also completes (cf. Propositions 1,2) the proof of

Theorem 1.

Our next objective is to prove that, for a large class of functions ω, $\mathcal{D}'_\omega$ is a PLAU-space (cf. Def. 3,I). Let $\mathfrak{M}_c = \{\omega \in \mathfrak{M} : \omega$ concave for $\xi_1 \geq 0, \ldots, \xi_n \geq 0$, and ω an even function in each variable separately$\}$. For each $\omega \in \mathfrak{M}_c$, let $\omega_1(t) = \omega(nt,0,\ldots,0)$, $\ldots,\omega_n(t) = \omega(0,\ldots,0,nt)$. Then

$$(54) \qquad \frac{1}{n}\sum_{j=1}^{n} \omega_j(\xi_j) \leq \omega(\xi) \leq \sum_{j=1}^{n} \omega_j(\xi_j) .$$

Since for each $\omega \in \mathfrak{M}_c$ and $\tilde{\omega} \overset{\text{def}}{=} \sum_{j=1}^{n} \omega_j$, the spaces $\mathcal{D}_\omega$ and $\mathcal{D}_{\tilde{\omega}}$ coincide, we shall usually replace each ω in $\mathfrak{M}_c$ by its modification $\tilde{\omega}$ which will be called ω.

If $P(t)$ is a decreasing convex function of $t \geq 0$, then for each $x,y \in \mathbb{R}^n$,

$$(55) \qquad P(|x|) + P(|y|) \leq P(|x| + |y|) + P(0) \leq P(|x+y|) + P(0).$$

Similarly, for Q concave and increasing on $[0,\infty)$,

$$(56) \qquad Q(|x+y|) + Q(0) \leq Q(|x|) + Q(|y|) \qquad (x,y \in \mathbb{R}^n)$$

Simple examples of functions ω of $\mathfrak{M}_c$ can be obtained by taking $Q_1,\ldots,Q_n$ arbitrary non-negative concave decreasing functions of $t \geq 0$, and setting

$$\omega(\xi_1,\ldots,\xi_n) = \sum_{j=1}^{n} P_j(|\xi_j|) .$$

Moreover, for Q as above, the function $\omega(\xi) = Q(|\xi|)$ is in $\mathfrak{M}_c$.

Theorem 2. For each $\omega \in \mathfrak{M}_c$, $\mathcal{D}'_\omega$ is a PLAU-space.[3]

Proof. First we have to verify condition (vii) of Def. 3,I. Since $\omega = \Sigma\omega_j$, it will follow that $\mathcal{D}'_\omega$ is the AU-product of 1-dimensional AU-spaces $\mathcal{D}'_{\omega_j}$, provided we can show that: (a) each $\mathcal{D}_{\omega_j}$ is nontrivial; and, (b) the AU- and BAU-structures of the spaces $\mathcal{D}'_{\omega_j}$ generate the corresponding structures for $\mathcal{D}'_\omega$ in the way described in Definition 3,I. To check (a), let us take, e.g., $j=1$, and set $S = \{(t_1,\ldots,t_n) : t_1 \geq 1,\ 2^{-1}t_1 \leq t_k \leq 2t_1\}$. Then, (cf. condition (β) at the beginning of this chapter)

$$(\log 4)^{n-1}\int_1^\infty \frac{\omega_1(t_1)dt_1}{t_1^2} = \left(\prod_{k=2}^{n}\int_{t_1/2}^{2t_1}\frac{dt_k}{t_k}\right)\int_1^\infty \cdots = \int_S \frac{\omega(nt_1,0,\ldots,0)}{t_1^2 t_2\cdots t_n}dt$$

$$\leq C_o \int_S \frac{\omega(nt)}{|nt|^{n+1}}\,dt \leq C_1 J_n(\omega) < \infty .$$

Thus $J_1(\omega_j) < \infty$ for all j, hence by [9], $\mathcal{D}_{\omega_j} \neq 0$. (Conversely, let ω_j be functions of one variable satisfying conditions (α),(β) for n=1. Then the function $\omega(\xi_1,\ldots,\xi_n) = \omega_1(\xi_1)+\ldots+\omega(\xi_n)$ also satisfies (α),(β). Indeed, if $\Omega(t) = \omega_1(t)+\ldots+\omega_n(t)$, then obviously $J_1(\Omega) < \infty$. However, for some constants C_2, C_3,

$$C_2 J_n(\omega) \leq \int_{|\xi|\geq 1} \frac{\omega(\xi_1)+\ldots+\omega_n(\xi_n)}{|\xi|^{n+1}}\,d\xi \leq \int_{|\xi|\geq 1}\frac{\Omega(|\xi|)}{|\xi|^{n+1}}\,d\xi \leq C_3 J_1(\Omega) \quad .)$$

(b) For any $\Lambda \in \mathcal{L}$, $\Lambda(t) = t\lambda(t)$, we have

$$\sum_{j=1}^{n} \omega_j \lambda(\omega_j) \leq \omega\lambda(\omega) \leq n \sum_{j=1}^{n} \omega_j \lambda(\omega_j) .$$

Therefore, if $\mathcal{M}_j$ denotes the family $\mathcal{M}(\mathcal{L}_c)$ for the space $\mathcal{D}'_{\omega_j}$, the family $\mathcal{M} \overset{\text{def}}{=} \{m(\zeta) = m_1(\zeta_1)\ldots m_n(\zeta_n) : m_j \in \mathcal{M}_j, \ 1 \leq j \leq n\}$ must be a BAU-structure for $\mathcal{D}'_\omega$. Let D_ω be the vector space $\bigotimes_{j=1}^{n} \mathcal{D}_{\omega_j}$ equipped with the topology $\mathcal{T}(D)$, defined as the unique bornological topology compatible with a fundamental system of bounded sets of the form $A(m,c) = \{\phi \in D : \sup_{\zeta \in \mathbb{C}^n} (|\hat{\phi}(\zeta)|/m(\zeta)) \leq c\}$. Obviously, the l.c. space D_ω is isomorphic to a subspace of $\mathcal{D}_\omega$. Thus the fact that the completion $\tilde{D}_\omega$ of D_ω is the whole space $\mathcal{D}_\omega$ follows from the density of D_ω in $\mathcal{D}_\omega$.[4] From the discussion preceding Def. 3,I we then obtain that there is an AU-structure $\mathcal{K}$ on $\mathcal{D}_\omega$ of the form required by condition (vi) of Def. 3,I; namely, we can define $\mathcal{K}$ as $\{k(\zeta) : k(\zeta) = k_1(\zeta_1)\ldots k_n(\zeta_n), k_j \in \mathcal{K}_j, \ \forall_j\}$.

To verify (vii) we have to limit ourselves to the spaces $\mathcal{D}_{\omega_j}$. Hence we fix j and call $\omega = \omega_j$, $\mathcal{M}(\mathcal{L}_c) = \mathcal{M}_j$, etc. Let $t = \nu(s)$ be the inverse function of $s = \omega(t)$ and

$$\mathcal{L}^* \overset{\text{def}}{=} \{\Lambda(r) = r\lambda(r) \in \mathcal{L}_c : \lambda(r)/\nu'(r) \to 0; \int_1^\infty t^{-2}\Lambda(\omega/t))dt < \infty\}.$$

It is clear that $\mathcal{M}(\mathcal{L}^*)$ is again a BAU-structure for $\mathcal{D}'_\omega$ (cf. the end of the proof of Prop. 3). Let us fix a Λ^* in $\mathcal{L}^*$. We claim that there exists a Λ in $\mathcal{L}$ such that

1. $\Lambda(\omega(t))$ is a concave function of $t \geq 0$ (hence $\Lambda(\omega(t)) \nearrow +\infty$);
2. $\Lambda(\omega(t)) \leq \Lambda^*(\omega(t)) + \text{const.}$

First, let us construct a continuous function $H(\delta)$ $(\delta \geq 0)$ for which $H(\delta) \to 0$, and

$$(57) \qquad \lambda^*(s) = \frac{\Lambda^*(s)}{s} \geq \frac{1}{s}\int_{\omega(0)}^{s} H(\delta)\nu'(\delta)d\delta \to \infty .$$

It suffices to take $H(s) = \min\{\lambda^*(\delta)/\nu'(\delta) : \omega(0) \leq \delta \leq s\}$. Then

$$\frac{1}{s}\int_{\omega(0)}^{s} H(\delta)\nu'(\delta)d\delta \leq \frac{1}{s}\int_{\omega(0)}^{s} \lambda^*(\delta)d\delta \leq \lambda^*(s) .$$

On the other hand, since the set $\{s : H(s) = \lambda^*(s)\nu'(s)\}$ cannot be bounded, $H(s)\nu'(s) \to \infty$, and (57) follows. Then, for $g(\tau) = H(\omega(\tau))$, we have

$$\int_{\omega(0)}^{s} H(\delta)\nu'(\delta)d\delta = \int_{0}^{t} g(\tau)d\tau .$$

Thus, by (57), we can set

$$\Lambda(\omega(t)) = \int_{0}^{t} g(\tau)d\tau + \text{const.}$$

Let $\mathcal{L}^{**}$ be the class of all such Λ's (i.e. Λ's constructed for all $\Lambda^* \in \mathcal{L}^*$). The family $\mathcal{M} = \mathcal{M}(\mathcal{L}^{**})$ is again a BAU-structure for $\mathcal{D}'_\omega$; and, we shall show that the class $\mathcal{M}$ satisfies condition (vii).

Given $m \in \mathcal{M}$ and $\varepsilon > 0$, we must first exhibit an $m' \in \mathcal{M}$ such that for any $z_o = x_o + iy_o \in \mathbb{C}$, there will be an entire function $\phi(z)$ for which

$$(58) \qquad \frac{m(z_o)|\phi(z)|}{\min\limits_{|\zeta - z_o| \leq \varepsilon} |\phi(\zeta)|} \leq m'(z) . \qquad (z \in \mathbb{C})$$

However, $m(\zeta) = C \exp[\chi(\xi) + A|\eta|]$ where $\chi(\xi) = -\Lambda(\omega(\xi))$, $\Lambda \in \mathcal{L}^{**}$.

It follows from the proof of Proposition 3 (cf. (53)) that $e^{\chi(\xi)} \in L^2$; by the definition of $\mathcal{L}^*$,

(59) $$\int_1^\infty \frac{|\chi(t)|}{t^2}\,dt < \infty .$$

By [42], Th. XII, there exists a function g with compact support, $g \neq 0$, and such that $|\hat{g}(\zeta)| \leq m(\zeta)$ for all ζ and $\hat{g}(\xi) \geq 0$. Since $\hat{g}(0) = \int_{-\infty}^{\infty} g(t)dt > 0$, there exist positive numbers δ, $\delta < \varepsilon$, and c such that, for $|\xi| \leq \delta$, $\hat{g}(\xi) > c$. Set

(60) $$m'(\zeta) = C_1\exp[6A|\eta| + \chi(\tfrac{\delta}{\varepsilon}\xi)];\; C_1 = c^{-1}C^2\exp[3A\varepsilon + \chi(0)] .$$

Given any $z_0 = x_0 + iy_0$, let $\alpha = -3A\text{sign}\, y_0$ and $\beta = -3A|y_0|$. We claim that the entire function $\phi(\zeta) = \hat{g}(\frac{\delta}{\varepsilon}(\zeta - z_0))\exp[i\alpha\zeta+\beta]$ satisfies (58). For $z = x+iy$,

(61) $$m(z_0)|\phi(z)| \leq C^2\exp[\chi(x_0) + A|y_0| + \chi(\tfrac{\delta}{\varepsilon}(x-x_0)) + \tfrac{\delta}{\varepsilon}A\,|y-y_0| + \mathrm{Re}(i\alpha z + \beta)] .$$

By (55) we have

(62) $$\chi(x_0) + \chi(\tfrac{\delta}{\varepsilon}(x-x_0)) \leq \chi(\tfrac{\delta}{\varepsilon}\,x_0) + \chi(\tfrac{\delta}{\varepsilon}(x-x_0)) \leq \chi(\tfrac{\delta}{\varepsilon}x) + \chi(0) .$$

Since $\mathrm{Re}(i\alpha z+\beta) \leq 3A|y| - 3A|y_0|$, estimates (61) and (62) give

(63) $$m(z_0)|\phi(z)| \leq C_2\exp[\chi(\tfrac{\delta}{\varepsilon}x) + 6A|y|]$$

where $C_2 = C^2e^{\chi(0)}$. Inequality (63) combined with the obvious estimate

$$\min_{|\zeta - z_0| \leq \varepsilon} |\phi(\zeta)| \geq ce^{-3A\varepsilon} \tag{64}$$

implies (60).

Now, for m, ε fixed as above, we have to find an $m' \varepsilon \mathcal{M}$ such that for each $z_0 = x_0 + iy_0$, one can find an entire function ψ, for which

$$|\psi(z)| \sup_{-\infty<\xi<\infty} \left\{ \frac{m(\xi+iy_0)}{\min\limits_{|r-y_0|\leq\varepsilon} |\psi(\xi+ir)|} \right\} \leq m'(z) \tag{65}$$

for all $z \in ¢$. Set $\Omega(\xi) = \chi(0) - \chi(\xi) + \log(1+\xi^2)$. Then the space $\mathcal{D}_\Omega(\mathbb{R})$ is well defined and $\mathcal{D}_\Omega(\mathbb{R}) \neq \{0\}$. Take an arbitrary function $g \in \mathcal{D}_\Omega$ such that $\operatorname{supp} g \subset (-A,A)$ and $\hat{g}(\xi) \geq 0$; and, let f be the function in $L^2(\mathbb{R})$ for which $\hat{f}(\xi) = \exp[\chi(\xi)]$. Let $h = f.g$. Then, for each μ, $0 < \mu < 1$,

$$|\hat{h}(\xi+i\eta)| \leq \int_{-\infty}^{\infty} |\hat{f}(t)\hat{g}(\xi-t+i\eta)|dt \leq C_\mu \int \exp[\chi(t)-\mu\Omega(\xi-t)+A|\eta|]dt$$

$$\leq C_\mu \exp[-\mu\Omega(\xi)+A|\eta|] \int \exp[\chi(t)+\mu\Omega(t)]dt \tag{66}$$

$$\leq \tilde{C}_\mu \exp[-\mu\Omega(\xi)+A|\eta|] \; .^{*}$$

In particular, this shows that $h \in \mathcal{D}_\omega$. On the other hand, by (55),

---

*If μ were ≥ 1, the last integral in (66) would not converge.

$$\hat{h}(\xi) = \int_{-\infty}^{\infty} \hat{f}(\xi-t)\hat{g}(t)dt = \int \exp[\chi(-t+\xi)]\hat{g}(t)dt$$

(67)

$$\geq e^{\chi(\xi)} \int \exp[\chi(-t) - \chi(0)]\hat{g}(t)dt = C_2 e^{\chi(\xi)} .$$

We need an estimate of $|\hat{g}(t+i\eta)-\hat{g}(t)| = |\eta||\hat{g}'(t+i\tilde{\eta})|$. Applying the Cauchy formula, we obtain for each $\mu > 0$,

$$(68) \quad |\hat{g}(t+i\eta)-\hat{g}(t)| \leq \tilde{C}|\eta| \max_{|u|\leq 1} |\hat{g}(t+i\eta+u)| \leq T_\mu |\eta| e^{A|\eta|-\mu\Omega(t)} .$$

Therefore, by (67) and (68),

$$|\hat{h}(\xi+i\eta)| = \left| \int \hat{f}(\xi-t)\hat{g}(t+i\eta)dt \right|$$

(69)

$$\geq \left| \int \hat{f}(\xi-t)\hat{g}(t)dt \right| - \int \hat{f}(\xi-t)\,|\hat{g}(t+i\eta) - \hat{g}(t)|dt$$

$$\geq C_2 e^{\chi(\xi)} - T_1 |\eta| e^{A|\eta|} \int \exp[\chi(\xi-t) - \Omega(t)]dt .$$

Using again the superadditivity of χ (cf. (55)), we get $\chi(\xi-t) - \Omega(t) \leq \chi(\xi) - \log(1+t^2)$. Then, for $|\eta|$ sufficiently small, say $|\eta| \leq \delta$ for some $\delta < \varepsilon$, (69) yields

$$(70) \quad |\hat{h}(\xi+i\eta)| \geq \frac{C_2}{2} e^{\chi(\xi)} .$$

Now, let us define m' as

$$(71) \quad m'(\zeta) = C_3 \exp[6A|\eta| - \tfrac{1}{2}\Omega(\tfrac{\delta}{\varepsilon}\xi)] \; ; \quad C_3 = 2CC_1C_2^{-1}e^{3A\varepsilon} \; ;$$

and, for $z_o = x_o+iy_o$ fixed, we set $\psi(\zeta) = \hat{h}(\frac{\delta}{\varepsilon}(\zeta-iy_o))\exp[i\alpha\zeta+\beta]$, where α,β are as above (see (60)). Then, by (70),

$$(72) \qquad |\psi(\xi+is)| = |\hat{h}(\tfrac{\delta}{\varepsilon}(\xi+i(s-y_o)))|\exp[3A(s-y_o)\operatorname{sign} y_o]$$

$$\geq \frac{C_2}{2}\exp[\chi(\tfrac{\delta}{\varepsilon}\xi) - 3A\varepsilon] ;$$

hence,

$$(73) \qquad \sup_{-\infty<\xi<\infty} \left\{2C_2^{-1}m(\xi+iy_o)\exp[3A\varepsilon - \chi(\tfrac{\delta}{\varepsilon}\xi)]\right\}$$

$$\leq C_4 e^{A|y_o|} ; \qquad C_4 = 2CC_2^{-1}e^{3A\varepsilon} .$$

Finally, by (73),

$$|\psi(z)| \sup_{-\infty<\xi<\infty} \{\ldots\}$$

$$\leq C_4 e^{A|y_o|} C_1\exp[-\tfrac{1}{2}\Omega(\tfrac{\delta}{\varepsilon}x) + A\tfrac{\delta}{\varepsilon}|y-y_o| + 3A(|y|-|y_o|)]$$

$$\leq m'(z),$$

and this completes the proof of Theorem 2.

§2. THE BEURLING SPACES $\mathcal{E}_\omega$, $\mathcal{E}'_\omega$.

From now on each ω will be assumed symmetric, i.e. $\omega(\xi) = \omega(-\xi)$. (We limit ourselves to such ω's only for the sake of simplicity, cf. [9].) $\mathfrak{M}_s$ will denote the subclass of $\mathfrak{M}$ containing all symmetric functions ω.

Definition 3. Let ω be a function in $\mathfrak{M}_s$. Then $\mathcal{E}_\omega$ is defined as the set of all functions ϕ on $\mathbb{R}^n$ such that, for each compact set K, the restrictions to K of ϕ and of some ψ in $\mathcal{D}_\omega$ agree.[5] The topology $\mathcal{T}(\mathcal{E}_\omega)$ is given by the system of seminorms $[\phi]_{\lambda,K}$ defined as

$$(74) \qquad [\phi]_{\lambda,K} = [\phi]^{(\omega)}_{\lambda,K} = \inf_{\psi=\phi \text{ in } K} |\psi|^{(\omega)}_\lambda$$

for all $\lambda > 0$ and all compact sets K. $\mathcal{E}'_\omega$ will denote the strong dual of $\mathcal{E}_\omega$.

Remarks: 5. The spaces $\mathcal{E}_\omega$, $\mathcal{E}'_\omega$ bear the same relationship to spaces $\mathcal{D}_\omega$, $\mathcal{D}'_\omega$ that the spaces $\mathcal{E}$, $\mathcal{E}'$ bear to the spaces $\mathcal{D}$, $\mathcal{D}'$. In particular, standard arguments show that $\mathcal{E}_\omega$ is a Fréchet-Montel space. Therefore $\mathcal{E}_\omega$, $\mathcal{E}'_\omega$ are reflexive, barreled, bornological, etc. Moreover, as a set, $\mathcal{E}'_\omega$ can be identified with the subspace of $\mathcal{D}'_\omega$, consisting of all elements with compact support. Hence, the elements of $\mathcal{E}'_\omega$ will be called the Beurling ω-distributions with compact support.

6. Paley-Wiener theorem for $\mathcal{E}'_\omega$: For each $\phi \in \mathcal{E}'_\omega$, there are constants $C > 0$, $A > 0$ and N real such that, for all $\zeta \in \mathbb{C}^n$,

(75) $$|\hat{\Phi}(\zeta)| \leq Ce^{N\omega(\xi)+A|\eta|} \quad ;$$

and, conversely, if g is an entire function satisfying (with some C,N,A as above) the last inequality, then $g \in \hat{\mathcal{E}}'_\omega$ [9].

7. Let K be a non-empty compact subset of $\mathbb{R}^n$. $\mathcal{E}'_\omega(K)$ will denote the subspace of $\mathcal{E}'_\omega$ defined by

$$\mathcal{E}'_\omega(K) = \{\Phi \in \mathcal{E}'_\omega : \text{supp } \Phi \subset K\}.$$

If $\{K_s\}_{s\geq 1}$ is a sequence of compact sets exhausting $\mathbb{R}^n$, then the bornologicity of $\mathcal{E}'_\omega$ implies (cf. [46]) that

$$\mathcal{E}'_\omega = \lim_{s \to \infty} \text{ind } \mathcal{E}'(K_s) \ .$$

8. Another system of norms on $\mathcal{E}_\omega$ can be defined as follows: for each $\lambda > 0$ and $\psi \in \mathcal{D}_\omega$, let

(76) $$\|\phi\|_{\lambda,\psi} = \|\phi\|^{(\omega)}_{\lambda,\psi} = |\psi\phi|^{(\omega)}_\lambda \quad .$$

Both systems (75) and (76) define the same topology. (Indeed, given $\lambda > 0$ and K compact, then for $\psi \equiv 1$ on K and $\psi \in \mathcal{D}_\omega$, $[g]_{\lambda,K} = [\psi g]_{\lambda,K} \leq \|g\|_{\lambda,\psi}$. Conversely, given $\lambda > 0$, $\psi \in \mathcal{D}_\omega$, let $K = \text{supp } \psi$. For each $\varepsilon > 0$ there is a $\phi \in \mathcal{D}_\omega$, $\phi = g$ on K and $|\phi|_\lambda \leq [g]_{\lambda,K} + \varepsilon$. Then $\psi\phi \equiv \psi g$ and $\|g\|_{\lambda,\psi} = |\psi\phi|_\lambda \leq |\psi|_\lambda |\phi|_\lambda \leq |\psi|_\lambda([g]_{\lambda,K} + \varepsilon)$, etc.). From here, it is easy to conclude that <u>the space $\mathcal{E}_\omega$ can be characterized as the space of all multipliers of the space $\mathcal{D}_\omega$</u>, i.e. as the space of all complex valued functions ϕ such that each mapping $M_\phi : \psi \mapsto \psi\phi$ is an endomorphism of the space $\mathcal{D}_\omega$. Thus $\mathcal{E}_\omega$ is a subspace of $L(\mathcal{D}_\omega,\mathcal{D}_\omega)$ and the topology of $\mathcal{E}_\omega$ is the (induced) topology of pointwise convergence in $L(\mathcal{D}_\omega,\mathcal{D}_\omega)$ (cf. [9]). For our purposes a similar description of the space $\mathcal{E}'_\omega$ will be needed:

Proposition 4. The space $\mathcal{E}'_\omega$ is the space of all convolutors of the space $\mathcal{D}_\omega$, i.e. $\mathcal{E}'_\omega$ consists of all distributions $\Psi \in \mathcal{D}'_\omega$ such that the mapping $\mathcal{C}_\Psi : \phi \mapsto \phi * \Psi$ is in $L(\mathcal{D}_\omega,\mathcal{D}_\omega)$. Moreover the topology of $\mathcal{E}'_\omega$ coincides with the compact open topology induced on $\mathcal{E}'_\omega$ from $L(\mathcal{D}_\omega,\mathcal{D}_\omega)$.[6]

Proof. Let $\Psi \in \mathcal{D}'_\omega$ be a convolutor of $\mathcal{D}_\omega$ such that $\operatorname{supp}\Psi$ is not compact. Let ρ_n be a regularizing sequence in $\mathcal{D}_\omega$, i.e. $\Psi_n = \Psi * \rho_n \to \Psi$ in $\mathcal{D}'_\omega$. We can assume that $\Psi_n \neq 0$ for all n. Let $\tilde{\Psi}_n = \Psi * \tilde{\rho}_n$ where $\{\tilde{\rho}_n\}$ is a new sequence in $\mathcal{D}_\omega$ defined as follows. First, let $\tilde{\rho}_1 = \rho_1$. Since Ψ is a convolutor of $\mathcal{D}_\omega$, there is an $r_1 > 0$ such that $\operatorname{supp}\tilde{\Psi}_1 \subset K_{r_1}$ (cf. notation in Def. 2,II). Let n_2 be so large that $\operatorname{supp}\Psi_{n_2} \underset{\neq}{\subset} K_{r_1}$; and let $x_1 \in \operatorname{supp}\tilde{\Psi}_1$. Define $\tilde{\rho}_2 = \gamma_2\rho_{n_2}$ where γ_2 is the constant defined by

$$\gamma_2 = \frac{|\tilde{\Psi}_1(x_1)|}{3^2\max[1,|\tilde{\Psi}_1(x_1)|,|\Psi_{n_2}(x_1)|]\max[1,\|\rho_{n_2}\|_1]} .$$

Now let $x_2 \in \operatorname{supp}\tilde{\Psi}_2 \setminus K_{r_1}$ and $\operatorname{supp}\tilde{\Psi}_2 \subset K_{r_2}$. Let n_3 be such that $\operatorname{supp}\Psi_{n_3} \underset{\neq}{\subset} K_{r_2}$ and $\tilde{\rho}_3 = \gamma_3\rho_{n_3}$ where

$$\gamma_3 = \frac{\min[|\tilde{\psi}_1(x_1)|,|\tilde{\psi}_2(x_2)|]}{3^3\max[1,|\tilde{\psi}_1(x_1)|,|\tilde{\psi}_2(x_2)|,|\psi_{n_3}(x_1)|,|\psi_{n_3}(x_2)|]\max[1,\|\rho_{n_3}\|_2]} ,$$

etc. Since $\operatorname{supp} \rho_n \subset K_{r_o}$ for all n, we also have $\operatorname{supp} \tilde{\rho}_n \subset K_{r_o}$ for all n. Let $\tilde{\rho} = \Sigma \tilde{\rho}_j$. This series converges in $\mathcal{D}_\omega$, because for all integers $m \geq 1$, $p \geq 1$, $N > m$,

$$\left\| \sum_{j=N}^{N+p} \tilde{\rho}_j \right\|_m \leq \sum_{j=N}^{N+p} \frac{1}{3^j} .$$

Hence $\tilde{\rho} \in \mathcal{D}_\omega$ and $\tilde{\psi} = \psi * \tilde{\rho} \in \mathcal{D}_\omega$. On the other hand, since $\tilde{\psi}_j(x_k) = 0$ for $j < k$, we get from the above construction of γ_j's,

$$|\psi(x_k)| \geq |\tilde{\psi}_k(x_k)| - \sum_{j>k} |\tilde{\psi}_j(x_k)|$$

$$\geq |\tilde{\psi}_k(x_k)| - \sum_{j>k} \frac{1}{3^j} |\tilde{\psi}_k(x_k)| > 0 .$$

Therefore $\operatorname{supp} \psi$ cannot be compact and this is a contradiction.

Let $\mathcal{T}_{c.o.}$ be the topology induced on $\mathcal{E}'_\omega$ by the compact open topology of $L(\mathcal{D}_\omega, \mathcal{D}_\omega)$. First, let us show that $\mathcal{T}_{c.o.}$ is coarser than $\mathcal{T}(\mathcal{E}'_\omega)$. By Remark 7 it is enough to show that, for any compact set K, the injection $\mathcal{E}'_\omega(K) \to (\mathcal{E}'_\omega, \mathcal{T}_{c.o.})$ is continuous. Let $\mathcal{W}$ be a $\mathcal{T}_{c.o.}$-neighborhood of the origin in $\mathcal{E}'_\omega$, i.e. for some bounded set B in $\mathcal{D}_\omega$ and a neighborhood $\mathcal{U}$ of the origin in $\mathcal{D}_\omega$, $\mathcal{W} = \mathcal{W}(B, \mathcal{U}) = \{\Phi \in \mathcal{E}'_\omega : \Phi * B \subset \mathcal{U}\}$. For a fixed K, there is a compact set K_1 such that, for $\Phi \in \mathcal{E}'_\omega(K)$ and $f \in B$, $\operatorname{supp}(\Phi * f) \subset K_1$. Moreover we can find $\lambda > 0$ and $\varepsilon > 0$ for which $\{\phi \in \mathcal{D}_\omega : \operatorname{supp} \phi \subset K_1, \ |\phi|_\lambda \leq \varepsilon\} \subset \mathcal{U}$. We need a bounded set A in $\mathcal{E}_\omega(K)$ such that, if $|\langle A, \Phi \rangle| \leq 1$, then $|\Phi * B|_\lambda \leq \varepsilon$. Set $A = \{g : \hat{g}(\xi) = \varepsilon^{-1} \hat{f}(\xi) e^{\lambda\omega(\xi)} \theta(\xi)$ where $f \in B$, and θ is an arbitrary measurable complex valued function such that $|\theta| \equiv 1\}$. (In fact, it would be sufficient to take for θ's only certain functions from $C^\infty(\mathbb{R}^n \smallsetminus M)$ where M is a "thin" set.) If $g \in A$

and $\psi \in \mathcal{D}_\omega$, then

$$|g\psi|_\mu = \int |\widehat{g\psi}(\xi)| e^{\mu\omega(\xi)} d\xi$$

$$\leq \iint |\hat{g}(\xi-t)\hat{\psi}(t)| e^{\mu\omega(\xi)} dt\, d\xi$$

$$= \varepsilon^{-1} \iint |\hat{f}(\xi-t)\hat{\psi}(t)| e^{\mu\omega(\xi)+\lambda\omega(\xi-t)} dt\, d\xi$$

$$\leq \varepsilon^{-1} \int \left[\int |\hat{f}(\xi-t)| e^{(\lambda+\mu)\omega(\xi-t)} d\xi \right] |\hat{\psi}(t)| e^{\mu\omega(t)}\, dt$$

$$= \varepsilon^{-1} |f|_{\lambda+\mu}\, |\psi|_\mu \ .$$

By Remark 8 this shows that A is bounded in $\mathcal{E}_\omega$ and hence also in $\mathcal{E}_\omega(K)$. Now, let us show that for $\Phi \in A^\circ$, $|\Phi * B|_\lambda \leq \varepsilon$. Actually, if $\Phi \subset A^\circ$ and $f \in B$, there is a $\theta(\xi)$ as above such that

$$|\Phi * f|_\lambda = \int |\hat{\Phi}(\xi)\hat{f}(\xi)| e^{\lambda\omega(\xi)} d\xi = \int \hat{\Phi}(\xi)\hat{f}(\xi)\theta(\xi) e^{\lambda\omega(\xi)} d\xi$$

$$= \varepsilon \int \hat{\Phi}(\xi)\hat{g}(\xi) d\xi = \varepsilon \langle g,\Phi\rangle \ .$$

Since $g \in A$, the result follows.

It remains to show that the topology $\mathcal{T}_{c.o.}(\mathcal{E}'_\omega)$ is finer than $\mathcal{T}(\mathcal{E}'_\omega)$. Let $\mathcal{U}$ be a neighborhood in $\mathcal{T}(\mathcal{E}'_\omega)$, i.e. $\mathcal{U} = B^\circ$ for some B bounded in $\mathcal{E}_\omega$. If α is an arbitrary lattice point in $\mathbb{R}^n$, let K_α be the cube of edge 2 and with center at α. Let $\{\psi_\alpha\}$ be a partition of unity subordinate to the covering $\{K_\alpha\}$. Let

$$\tilde{B} = \{2^{|\alpha|+n}\, \psi_\alpha f : V(f \in B;\alpha)\} \ .$$

As can be easily seen by Remark 8, the set $\tilde{B}$ is bounded in $\mathcal{E}_\omega$. For each $f \in B$, we can write $f = \sum_\alpha 2^{-|\alpha|-n} f_\alpha$, where $f_\alpha \in \tilde{B}$. Therefore $\tilde{B}^o \subset B^o$. We want to find a bounded subset A in $\mathcal{D}_\omega$ and a neighborhood $\mathcal{V}$ of the origin in $\mathcal{D}_\omega$ such that $\mathcal{W}(A,\mathcal{V}) \subset \tilde{B}^o$. Let Φ_α be functions in $\mathcal{D}_\omega$ such that $\Phi_\alpha \equiv 1$ on K_α. (Therefore, $\|g\|_{\Phi_{\alpha,\lambda}} = |g|_\lambda$ if $\operatorname{supp} g \subset K_\alpha$.) Since $\tilde{B}$ is bounded, there are constants $C_{\alpha,\lambda}$ such that $\|g\|_{\Phi_{\alpha,\lambda}} \leq C_{\alpha,\lambda}$ for all $g \in \tilde{B}$. Let $C_\lambda = \max_{|\alpha| \leq \lambda} C_{\alpha,\lambda}$. Let us choose positive numbers δ_α such that $C_{\alpha,\lambda}\delta_\alpha \leq C_\lambda$ for all λ,α. Then A is defined as the set of all $\phi \in \mathcal{D}_\omega$ such that $\operatorname{supp} \phi \subset K_0$ ($= K_\alpha$ for $\alpha = 0$) and $|\phi|_\lambda \leq C_\lambda$ for all $\lambda > 0$. Denote by $\tau_{-\alpha}$ the translation $\phi(x) \mapsto \phi(x+\alpha)$ and $\mathcal{V} = \{\phi \in \mathcal{D}_\omega : \max_{x \in K_\alpha} |\phi(x)| \leq \delta_{-\alpha}\}$. Then $\mathcal{V}$ is obviously a convex, closed and absorbing subset of $\mathcal{D}_\omega$; hence, $\mathcal{V}$ is a neighborhood in $\mathcal{D}_\omega$. Let $\mathcal{W} \overset{\text{def}}{=} \mathcal{W}(A,\mathcal{V})$. Then, for each $f \in \tilde{B}$, $f = \delta_\alpha^{-1}\tau_{-\alpha}(\phi)$ for some α and $\phi \in A$. Therefore, if $S \in \mathcal{W}$, then $|\langle f,S\rangle| = \delta_\alpha^{-1}|(S * \check{\phi})(-\alpha)| \leq \delta_\alpha^{-1}\varepsilon_{-\alpha} = 1.$* This shows that $\mathcal{W} \subset B^o$ and Proposition 4 is proved.

Let $\mathcal{K} = \mathcal{K}(\mathcal{E}_\omega)$ be the family of all functions k constructed as follows. Let $\mathcal{C}_0 = \mathcal{C} \setminus \mathcal{C}_+$. (For the definition of classes $\mathcal{C}$ and $\mathcal{C}_+$, cf. the text preceding Lemma 2.) Furthermore, let h be an increasing function on the real line such that $\inf h(s) > -\infty$, and the restriction of h to $[0,\infty)$ is in $\mathcal{C}_0$. Next we pick a function $\varepsilon(s)$ defined for s real, $0 < \varepsilon(s) \leq 1$, and so rapidly decreasing to 0 when $s \to -\infty$ that the function $\rho_\varepsilon(s)$, defined as the inverse function of $-\log \varepsilon(-s)$ $(s \geq 0)$, is in $\mathcal{C}$. Finally let μ be an arbitrary positive number. Then the series

$$(77) \quad k(\zeta) = k(h;\varepsilon;\mu;\zeta) = \sum_{s=-\infty}^{\infty} \varepsilon(s)\exp[|\eta|h(s) - (s+\mu)\omega(\xi)]$$

---

*If $\phi \in A$, then also $\check{\phi}(\xi) = \phi(-\xi) \in A$.

is locally uniformly convergent in $\mathbb{C}^n$. $\mathcal{K}(\mathcal{E}_\omega)$ will denote the family of all such majorants k.

Let $\mathcal{M} = \mathcal{M}(\mathcal{E}_\omega)$ be the class of all functions $m(\zeta) = Ce^{N\omega(\xi)+A|\eta|}$ (cf. (75)).

Theorem 3. The space $\mathcal{E}_\omega$ is an AU-space with basis $\mathcal{E}'_\omega$, an AU-structure $\mathcal{K}(\mathcal{E}_\omega)$ and a BAU-structure $\mathcal{M}(\mathcal{E}_\omega)$.

Proof. To begin with, let us first remark that conditions (i), (ii), (iv), (v) and (vi) are obvious. Thus it suffices to prove (A) and (B) of condition (iii). Let F be an entire function such that for each $k \in \mathcal{K}(\mathcal{E}_\omega)$,

$$(78)\qquad |F(\zeta)| \leq Ck(\zeta) = C\left(\sum_{s=-\infty}^{-1} + \sum_{s=0}^{\infty}\right) = C(\Sigma_- + \Sigma_+)$$

for some $C > 0$ and all $\zeta \in \mathbb{C}^n$. By (38),

$$\Sigma_+ \leq (5 + 2|\eta|)e^{|\eta|h(|\eta|)}$$

$$\Sigma_- = \sum_{\delta=1}^{\infty} \exp[\log \varepsilon(-\delta) + |\eta|h(-\delta) + (\delta-\mu)\omega(\xi)]$$

$$\leq \sum_{\delta=0}^{\infty} \exp[-\rho_\varepsilon(\delta)+\delta\omega(\xi)] \leq (5 + 2\omega(\xi))e^{\omega(\xi)\rho_\varepsilon(\omega(\xi))}.$$

Therefore,

$$(79)\qquad |F(\zeta)| \leq \tilde{C}(1+\omega(\xi))(1+|\eta|)\exp[(\omega(\xi)+|\eta|)(\rho_\varepsilon(\omega(\xi))+h(|\eta|))].$$

Let

$$G(\zeta) \overset{\text{def}}{=} \log\{|F(\zeta)|(1+\omega(\xi))^{-1}(1+|\eta|)^{-1}\}/(\omega(\xi)+|\eta|) .$$

The function G is bounded. Indeed, if it were not so, then for some ζ_n, $|\zeta_n| \to \infty$ and $|G(\zeta_n)| \geq n^2$. We can assume that $|\xi_n| \nearrow \infty$ and $|\eta_n| \nearrow \infty$. (The remaining cases are even simpler.) Let $\tilde{\rho}, \tilde{h}$ be the functions obtained by the linear interpolation of the values $\tilde{\rho}(0) = 1$, $\tilde{\rho}(\omega(\xi_n)) = n$ $(n \geq 1)$ and $\tilde{h}(0) = 1$, $\tilde{h}(|\eta_n|) = n$, respectively. Then we find functions $h^* \in \mathcal{C}_o$, $h^* \leq h$ and $\rho^* \in \mathcal{C}$, $\rho^* \leq \tilde{\rho}$, and apply (79) to the majorant $k = k(h^*;\ldots)$

$$1 \leq \frac{|G(\zeta_n)|}{n^2} \leq \frac{\log \tilde{C}}{n^2} + \frac{2}{n} \to 0$$

which is a contradiction. The rest follows similarly as in the proof of (47).

It remains to verify property (B). Let $k(\zeta) = k(h;\varepsilon;\mu;\zeta)$ be a fixed majorant in $\mathcal{K}$. We claim that the set $\mathcal{V}(k) = \{\phi \in \mathcal{E}'_\omega : |\hat{\phi}(\zeta)| \leq k(\zeta)\ (\forall\ \zeta)\}$ is a barrel in $\mathcal{E}'_\omega$. By Remark 5 this will imply that $\mathcal{T}(\mathcal{K}(\mathcal{E}'_\omega))$ is coarser than $\mathcal{T}(\mathcal{E}'_\omega)$. Each $\hat{\phi}$, $\phi \in \mathcal{E}'_\omega$, satisfies inequality (75) with some constants C,A,N. However, for each $\ell > 0$,

$$e^{N\omega+A|\eta|} \leq e^{2N\omega+\ell\omega-M|\eta|} + e^{2A|\eta|-\ell\omega+M|\eta|} = E_1 + E_2 ,$$

where $M \overset{\text{def}}{=} \inf h$. Let ℓ be so large that $h(\ell-\mu) \geq 2A+M$. Then, for $s = -(2N+\ell)$,

$$E_1 \leq e^{-s\omega+|\eta|h(s)} \leq \frac{k}{\varepsilon(s)} ,$$

and

$$E_2 \leq e^{|\eta|h(\ell-\mu)-\ell\omega} \leq \frac{k}{\varepsilon(\ell-\mu)} .$$

This shows that $\mathcal{V}(k)$ absorbs the distribution ϕ.

Now we have to show that $\mathcal{T}(\mathcal{K}(\mathcal{E}'_\omega))$ is finer than $\mathcal{T}(\mathcal{E}'_\omega)$. By Proposition 4, this is the same as showing that $\mathcal{T}(\mathcal{K}(\mathcal{E}'_\omega))$ is finer than $\mathcal{T}_{c.o.}$ Let $\mathcal{W} = \mathcal{W}(B,\mathcal{U})$ be a neighborhood of the origin in the latter topology. We can assume that B is defined by a function $m \in \mathcal{M}(\mathcal{L}_c)$ (cf. Proposition 3) and $\mathcal{U}$ is defined by means of some $\tilde{k} \in \mathcal{K}(\omega)$ (cf. (7)). If we can find a $k \in \mathcal{K}(\mathcal{E}_\omega)$ such that $km \leq \tilde{k}$, the proof will follow. Let us write $\tilde{k} = \tilde{k}(\tilde{h};\tilde{\varepsilon};\tilde{\mu};\zeta)$, $m = m(\Lambda;A;1;\zeta)$ and $\Lambda(t) = t\lambda(t)$; we want a $k = k(h;\varepsilon;\mu;\zeta)$ such that

$$(80)\quad \begin{cases} k(\zeta) = (\Sigma_- + \Sigma_+)\varepsilon(s)\exp[|\eta|h(s) - (s+\mu)\omega(\xi)] \\ \leq \sum_{s=1}^{\infty} \tilde{\varepsilon}(s)\exp[|\eta|(\tilde{h}(s)-A) + \omega(\xi)(\lambda(\omega(\xi)) - s - \tilde{\mu})] . \end{cases}$$

For all s large, say $s \geq s_A$, $\tilde{h}(s) \geq A$. For $s \geq 0$ we choose $h(s) \leq \tilde{h}(s+s_A)-A$, $\varepsilon(s) = \frac{1}{2}\tilde{\varepsilon}(s+s_A)$ and $\mu \geq \tilde{\mu}+s_A$. Setting $s+s_A = \delta$ we then obtain

$$(81)\quad \Sigma_+ \ldots \leq \frac{1}{2}\sum_{\delta \geq s_A} \tilde{\varepsilon}(\delta)\exp[|\eta|(\tilde{h}(\delta)-A) - (\delta+\tilde{\mu})\omega(\xi)] \leq \frac{1}{2}k(\zeta) .$$

It suffices to complete the definition of h and ε for $s < 0$ so that

$$(82)\quad \Sigma_- \ldots \leq \tilde{\varepsilon}(s_A)\exp[\omega(\xi)(\lambda(\omega(\xi)) - s_A - \tilde{\mu})] .$$

Let $h(s) = 0$ for $s < 0$ and $\varepsilon(s)$ $(s < 0)$ such that $\rho_\varepsilon \leq \tilde{\lambda}$. Then, by Lemma 2,

$$\Sigma_- \leq e^{-\mu\omega(\xi)} \sum_{s=-\infty}^{-1} e^{\log\varepsilon(s)-s\omega(\xi)} \leq e^{-\mu\omega(\xi)}(5+2\omega(\xi))\exp[\omega(\xi)\rho_\varepsilon(\omega(\xi))] \; .$$

For μ sufficiently large, the last inequality implies (82); this together with (81) proves inequality (80) and thus also the theorem.

Remarks: 9. Comparing Theorem 1 to Theorem 3 we see that there is an interesting relationship between the spaces $\mathcal{D}_\omega$ and $\mathcal{E}'_\omega$. The family $\mathcal{K}(\omega)$, which is the AU-structure for $\mathcal{D}'_\omega$ discussed in the previous section, can be obtained by taking the "Taylor" parts of all $k \in \mathcal{K}(\mathcal{E}_\omega)$; i.e., $\mathcal{K}(\mathcal{D}'_\omega)$ is comprised of all functions $\tilde{k}$ where $\tilde{k} = \Sigma_+$ for some $k = \Sigma_- + \Sigma_+ \in \mathcal{K}(\mathcal{E}_\omega)$. Moreover, from here it would not be difficult to derive another relationship between $\mathcal{D}_\omega$ and $\mathcal{E}'_\omega$; roughly, it can be described by saying that "outside a certain neighborhood of the real subspace of $\mathbb{C}^n$, the topologies of $\hat{\mathcal{D}}_\omega$ and $\hat{\mathcal{E}}'_\omega$ are the same" (cf. [6]). Namely, let $\alpha \in \mathcal{C}$, $\omega(\xi)/\alpha(|\xi|) \to 0$ for $|\xi| \to \infty$ and $R_\alpha \overset{\text{def}}{=} \{\zeta : |\eta| \geq \alpha(|\xi|)\}$. Then, for each $k \in \mathcal{K}(\omega)$, there is a majorant $k_1 \in \mathcal{K}(\mathcal{E}_\omega)$ such that $0 < C_1 \leq k_1(\zeta)/k(\zeta) \leq C_2$ in R_α. This statement generalizes to Beurling spaces a result of L. Ehrenpreis [19,23]. It also has analogous consequences for the study of hypoellipticity.

10. There are other variations of describing the topology of $\mathcal{E}'_\omega$ by means of series (77). Thus, for instance, one way of defining the topology of $\mathcal{E}'_\omega$ is the following. For each $k \in \mathcal{K}(\mathfrak{D}_\omega)$, let $\mathcal{V}(k)$ be the set of all Φ in $\mathcal{E}'_\omega$ such that, for some constants N and C (depending both on Φ),

(83) $$|\hat{\Phi}(\zeta)| \leq Ce^{N\omega(\xi)}k(\zeta)$$

for all ζ. The system $\{\mathcal{V}(k)\}$ defines a basis of neighborhoods in $\mathcal{E}'_\omega$ [6,7]. (The constant N in (83) is not necessarily a positive number, and is related to the order of the distribution Φ.)

11. It is very likely that under the same restrictions on ω's as in the previous section (cf. Theorem 2), one could prove that $\mathcal{E}_\omega$ is a PLAU-space.

CHAPTER III

Spaces of Approximate Solutions to Certain Convolution Equations

§1. SPACES $\mathcal{E}_B(L;\Phi)$

In this chapter we shall study another class of function spaces which are closely related to certain convolution equations. Roughly speaking, $\mathcal{E}_B(L;\Phi)$ will be the set of all "approximate solutions" of a convolution equation, which satisfy certain "growth conditions".

Let $\Phi: \mathbb{R}^n \to [0,+\infty]$ be a convex function such that $\Phi(0) = 0$, $\Phi(x_1,\dots,-x_j,\dots,x_n) = \Phi(x_1,\dots,x_j,\dots,x_n)$ for any j, and $\Phi(x)/|x| \to \infty$ for $|x| \to \infty$. Let $B = \{b_j\}$ be a convex p-sequence, i.e. $b_j = \exp(g(j))$, where $g: \mathbb{R}^p \to \mathbb{R}_+$ is a convex function such that $g(x)/|x| \to \infty$ when $|x| \to \infty$. (In particular, $b_j^{1/|b_j|} \to \infty$ when $j \to \infty$.) Finally, let $L = (L_1,\dots,L_p)$ be a vector with components $L_i \in \mathcal{E}'(\mathbb{R}^n)$, $1 \le i \le p$. If $j = (j_1,\dots,j_p)$ is a multiindex, we set

$$\hat{L}^j(z) = \prod_{i=1}^{p} (\hat{L}_i(z))^{j_i}.$$

<u>Definition 1</u>. For B, L, Φ as above, $\mathcal{E}_B(L;\Phi)$ is defined as the space of all C^∞-functions on $\mathbb{R}^n$ such that, for any $\varepsilon > 0$ and any multi-index α, there is a constant $C = C(f,\varepsilon,\alpha)$ so that, for all $x \in \mathbb{R}^n$ and $j = (j_1,\dots,j_p)$,

$$(1) \qquad |D^\alpha(L^j * f)(x)| \le C\varepsilon^{|j|} b_j \exp(\Phi(\varepsilon x)) .$$

Let

$$q_{\alpha,\varepsilon}(f) \overset{\text{def}}{=} \sum_j \frac{1}{b_j \varepsilon^{|j|}} \sup_x \frac{|D^\alpha(L^j * f)(x)|}{\exp[\Phi(\varepsilon x)]}$$

In the topology generated by the seminorms $q_{\alpha,\varepsilon}$, $\mathcal{E}_B(L;\Phi)$ becomes a Fréchet-Montel space.[1]

Remark 1. For $L = (\frac{\partial}{\partial x_1},\dots,\frac{\partial}{\partial x_n})$, the classes just defined are (roughly speaking) the Denjoy-Carleman classes and the Gevrey classes [38,42]. The α's and ε's are introduced so that we obtain (FM)-spaces.

Let Φ^* be the Young conjugate of Φ (cf. [53]), i.e.

(2) $$\Phi^*(y) = \max_x (\langle x,y\rangle - \Phi(x)) ,$$

and λ the series

(3) $$\lambda(z) = \sum_j \frac{|z|^{|j|}}{b_j}$$

which is convergent for all z. Finally, let us recall that W is called a weak AU-space, provided W satisfies conditions (i), (ii), (iii) of Def. 1, I with some AU-structure $\mathcal{K}$ such that $\hat{S} = \mathcal{O}(k)$ for each $S \in U$ and $k \in \mathcal{K}$ (cf. Remark 6 in Chap. I, §2).

Theorem 1. $\mathcal{E}_B(L;\Phi)$ is a weak AU-space with an AU-structure $\mathcal{K}$ containing all functions $k(z)$ on $\mathbb{C}^n$ such that for arbitrary constants N, c, $d > 0$, k satisfies the estimate

(4) $$(1+|z|)^N \lambda(c\hat{L}(z)) \exp [\Phi^*(d\cdot \operatorname{Im} z)] = \mathcal{O}(k(z)) .$$

Denote by $\Theta(N;c;d;z)$ the function in the left-hand side of (4). Moreover, every entire function $F(z)$ which is bounded by some function $\Theta(N;c;d;z)$ is an element of $\hat{\mathcal{E}}_B'(L;\Phi)$.[2]

Proof: For the sake of simplicity we shall give the proof only for $p = 1$, i.e. for a single convolution operator. The general case can be proved along the same lines. We set $U = \mathcal{E}_B'(L;\Phi)$ and $W = \mathcal{E}_B(L;\Phi)$. First we have to check that all exponentials $e^{i\langle x,z\rangle}$ lie in W. Indeed, for any fixed z, α and ε, we have by (4),

(5) $$q_{\alpha,\varepsilon}(e^{i\langle x,z\rangle}) \leq \Theta(|\alpha|;c;d;z) < \infty ,$$

where c and d depend only on ε. If T is an arbitrary element of $\mathcal{E}'_B(L;\Phi)$, T must be bounded on some neighborhood $\mathcal{N} = \mathcal{N}(q_{\alpha,\varepsilon})$; in particular, applying T to $e^{i<x,z>}$, we obtain from (5) that for some C (independent of z),

$$|\hat{T}(z)| \leq C\Theta(|\alpha|;c;d;z) \quad . \tag{6}$$

Now we are going to prove that the Fourier transform defines an isomorphism of the vector spaces U and V, where by V we denoted the vector space of all entire functions satisfying estimates of the form (6).* This will also prove that the set of exponentials $e^{i<x,z>}$ is total in $\mathcal{E}_B(L;\Phi)$; and, furthermore that $\mathcal{K}$ has property (A). First we must establish an intrinsic description of the space U.

Lemma 1. For each T ε U there exist positive constants ε,A,N and entire functions $Q_j(z)$ satisfying

$$\hat{T}(z) = \sum_{j=0}^{\infty} Q_j(z)\, \hat{L}^j(z) \tag{7}$$

and

$$|Q_j(z)| \leq A(1+|z|)^N \exp\, [\Phi^*(\tfrac{1}{\varepsilon}\, \mathrm{Im}\, z)]/\varepsilon^j b_j \quad . \tag{8}$$

Conversely, if F(z) is an entire function which can be expanded as in (7) with coefficients satisfying (8), then F = $\hat{T}$ for some T ε U.

Proof: Let us recall that the space $\mathcal{E}(\Phi)$ (cf. [4,23]) is the space of C^∞-functions satisfying conditions (1) with L = identity and $b_j=\infty$ for all $j \geq 1$ and $b_o = 1$. The space $\mathcal{E}(\Phi)$ is equipped with the natural topology defined by the seminorms

$$|g|_{\alpha,\varepsilon} = \sup_x \frac{|D^\alpha g(x)|}{\exp\, [\Phi(\varepsilon x)]} \quad .$$

*When this has been proved we shall, of course, write V = $\hat{U}$.

Every $f \in W$ can be mapped onto a sequence $\{f_j\} = \tilde{f}$ of functions in $\mathcal{E}(\Phi)$ by means of the mapping $f_j = L^j * f$, $j \geq 0$; and, any such sequence $\tilde{f}$ satisfies, for all α,ε,

$$p_{\alpha,\varepsilon}(\tilde{f}) = \sum_j (|f_j|_{\alpha,\varepsilon}/\varepsilon^j b_j) < \infty \ . \tag{9}$$

If $\tilde{W}$ is the space of all sequences $\tilde{f}$, then the seminorms (9) define a Fréchet topology on $\tilde{W}$; and, obviously $f \mapsto \tilde{f}$ is an isomorphism of l.c spaces W and $\tilde{W}$. This isomorphism shows that with every $T \in U$ we can associate a (not necessarily unique) sequence $\{T_j\}$, $T_j \in \mathcal{E}'(\Phi)$, such that

$$< T,f > = \sum_j < T_j, f_j > \ ;$$

and, for some r,ε, $A > 0$ and for all j and $g \in \mathcal{E}(\Phi)$,

$$|< T_j, g>| \leq A \max \{|g|_{\alpha,\varepsilon} : |\alpha| \leq r\} / \varepsilon^j b_j \ .$$

Conversely, any such sequence $\{T_j\}$ defines an element T in U by the formula

$$< T,f > = \sum_j < T_j, L^j * f > \ , \tag{10}$$

and the lemma follows if we set $Q_j = \hat{T}_j$. (For the characterization of $\mathcal{E}'(\Phi)$, see [4,23].)

(Proof of Theorem 4 continued) 1. <u>$T \to \hat{T}$ is injective</u>: Assume $\hat{T}(z) \equiv 0$. By Lemma 1 we can write

$$\hat{T}(z) = \sum_j Q_j(z)\, \hat{L}^j(z) \ .$$

Let $H(z,w) \overset{\text{def}}{=} \sum_j Q_j(z) w^j$ for $z \in \mathbb{C}^n$ and $w \in \mathbb{C}$. Then H is an entire function of $(z,w) \in \mathbb{C}^{n+1}$, for the coefficients are entire and satisfy the estimates (8). Moreover,

$$|H(z,w)| \leq A(1+|z|)^N \lambda(\tfrac{w}{\varepsilon}) \exp [\hat{\Phi}^*(\tfrac{1}{\varepsilon} \operatorname{Im} z)]$$

for some $A,N,\varepsilon > 0$ and all z. Since $H(z,\hat{L}(z)) = \hat{T}(z) \equiv 0$, the function

(11) $$G(z,w) = \frac{H(z,w)}{w - \hat{L}(z)}$$

is entire, and $|G(z,w)| \leq 4 \max_{|w-w'|\leq 1} |H(z,w')|$ (cf. Lemma 2, IV). Thus the function $G(z,w)$ can be written as a power series in w,

(12) $$G(z,w) = \sum_{j=0}^{\infty} G_j(z) w^j .$$

We shall show that the entire functions $G_j(z)$ satisfy uniform estimates of the form (8). Since we can always find $x \geq 1$ such that

$$\max_{k\geq 0} \left\{\frac{x^k}{b_k}\right\} = \frac{x^j}{b_j} ,$$

we obtain

$$\inf \{\lambda(\tfrac{w'}{\varepsilon})/|w|^j : w \in ₵, |w-w'| \leq 1\} \leq \frac{2^{2j+1}}{\varepsilon^j b_j} .$$

Then the Cauchy estimates yield

$$|G_j(z)| \leq 2A(1+|z|)^N \exp [\Phi^*(\tfrac{\operatorname{Im} z}{\delta})]/\delta^j b_j \qquad (\delta = \tfrac{\varepsilon}{4}) .$$

Comparing coefficients of equal powers of w in (11) (cf. (12) and the definition of $H(z,w)$) we obtain

$$\hat{T}_0 = - G_0\hat{L} ,$$

$$\hat{T}_j = G_{j-1} - G_j\hat{L} \qquad (j \geq 1)$$

If $S_j \in \mathcal{E}'(\Phi)$ is defined by $\hat{S}_j = G_j$, then $G_j\hat{L} = (L*S_j)^\wedge$ and $< S_j*L,g> = < S_j,L*g>$ for every $g \in \mathcal{E}(\Phi)$. Substituting the above identities in (10) we get, for every $f \in W$,

$$\begin{aligned} \langle T,f\rangle &= \sum_{j=0}^{\infty} < T_j,L^j*f> \\ &= - \langle S_0,L*f\rangle + \sum_{j=1}^{\infty} (\langle S_{j-1},L^j*f\rangle - < S_j,L^{j+1}*f>) = 0. \end{aligned}$$

Hence $T = 0$ and the injectivity of the mapping $T \mapsto \hat{T}$ is proved.

2. $T \mapsto \hat{T}$ is surjective. The proof will be based on the same idea used in the proof of the injectivity of $T \mapsto \hat{T}$. The main point consists in finding an entire function $H(z,w)$ satisfying good estimates and such that $H(z,\hat{L}(z)) = F(z)$ where F is an arbitrary fixed element of V; or, in other words, given a function F analytic on the variety $\{w-\hat{L}(z) = 0\}$ in $\mathbb{C}^{n+1}$, we are supposed to extend F to an entire function in $\mathbb{C}^{n+1}$ which would still have good bounds. This, however, is a typical problem to which the L^2-estimates of the $\bar{\partial}$-operator (cf. [23], Theorem 4.4.3) can be applied.

Let ρ be a C^∞-function in $\mathbb{R}^2 = \mathbb{C}$ such that $0 \leq \rho(s) \leq 1$ for all $s \in \mathbb{C}$; $\rho = 1$ for $|s| \leq 1/2$; $\rho = 0$ for $|s| \geq 1$; and, for some constant $C \geq 0$, $|\frac{\partial\rho}{\partial\bar{s}}| \leq C$. If F is an entire function in $\mathbb{C}^n$ such that

$$(13) \qquad |F(z)| \leq A\Theta(N;\alpha;\beta;z) ,$$

H will be defined by

$$H(z,w) = F(z)\rho(\omega-\hat{L}(z)) + (\omega-\hat{L}(z))u(z,w) .$$

Clearly, $H(z,\hat{L}(z)) = F(z)$ and we have to find the function u so that H is entire, i.e., $\bar{\partial}H = 0$, or

$$(14) \qquad \bar{\partial}u = \frac{-F(z)\bar{\partial}[\rho(w-\hat{L}(z))]}{w - \hat{L}(z)} .$$

(This expression is well defined, for the numerator vanishes when $|w-\hat{L}(z)| < 1/2$.) By virtue of the Paley-Wiener theorem, there are constants $D \geq 1$, $M > 0$ and $B > 0$ such that, for all z,

$$(15) \qquad \max\left\{\left|\frac{\partial\hat{L}(z)}{\partial z_j}\right|, |\hat{L}(z)|\right\} \leq D(1+|z|)^M \exp(B|\mathrm{Im}\, z|) .$$

Let us set $\zeta = (z,w)$, $\tilde{C} = 2(n+1)DC$ and (cf. [29])

$$|\bar\partial u|^2 = |\frac{\partial u}{\partial \bar w}|^2 + \sum |\frac{\partial u}{\partial \bar z_j}|^2$$

Then, using the fact that expression (14) vanishes for $|w-\hat{L}(z)| > 1$, we derive from (13), (14) and (15) the inequality

$$(16) \quad |\bar\partial u|^2 \le [\tilde{C}(1+|z|)^{M+N} \lambda(\alpha|w|+\alpha) \exp (B|\mathrm{Im}\, z|+\Phi^*(\beta\, \mathrm{Im}\, z))]^2$$

$$= \exp (\phi(\zeta)).$$

The function ϕ defined by (16) is plurisubharmonic. Hence, by Theorem 4.4.3 in [29], there exists a solution u of (14) satisfying

$$2 \int |u(\zeta)|^2 e^{-\phi(\zeta)} (1+|\zeta|^2)^{-(n+3)} |d\zeta| \le \int (1+|\zeta|^2)^{-(n+1)} |d\zeta| \overset{\text{def}}{=} \kappa$$

Therefore H is entire and

$$(17) \quad \int |H(\zeta)|^2 e^{-\phi(\zeta)} (1+|\zeta|^2)^{-(n+3)} (1+|w|^2+|\hat{L}(z)|^2)^{-1} |d\zeta| \le 6\kappa .$$

Using Lemma 1, I, we obtain from (17) the estimate

$$(18) \qquad |H(\zeta)| \le \kappa_0 (1+|z|)^p \lambda(2\alpha|w|) \exp (\Phi^*(\tilde{\beta}\, \mathrm{Im}\, z))$$

where $p = 2M+N+2n+6$, $\tilde{\beta} = 2(\beta+B)$ and κ_0 is some positive constant. Expanding $H(\zeta)$ into the power series, $H(\zeta) = \sum H_j(z)w^j$, we see that by (18),

$$(19) \qquad |H_j(\zeta)| \le \kappa_0 (1+|z|)^p \exp (\Phi^*(\tilde{\beta}\, \mathrm{Im}\, z))/\varepsilon^j b_j ,$$

where $\varepsilon = 1/4\alpha$. The surjectivity of the mapping $T \mapsto \hat{T}$ then follows by Lemma 1, because we can write

$$F(z) = H(z,\hat{L}(z)) = \sum H_j(z)\, \hat{L}^j(z) .$$

Finally, it is clear that V can be viewed as the l.c. space $\mathcal{A}(\mathcal{K})$ with $\mathcal{K}$ described in the statement of Theorem 4. We claim

3. <u>$T \mapsto \hat{T}$ is an isomorphism of l.c. spaces U and V</u>. Actually, since U is reflexive, U is also barreled; hence, $T \mapsto T$ is continuous. On the other hand, it can be shown [49] that V is the inductive limit of the Banach spaces $\mathcal{B}(N)$,

$$\mathcal{B}(N) = \{F \in \mathcal{A} : |F(z)| = \mathcal{O}(\Theta(N;N;N;z))\} .$$

In particular, V is a bornological space. However all constants in (19) depend only the constants occurring in (13), but not on the function F itself. Therefore, the mapping $\hat{T} \mapsto T$ maps bounded sets into bounded sets, and thus it is continuous. This completes the proof of Theorem 1.[3]

§2. A UNIQUENESS THEOREM FOR CONVOLUTION EQUATIONS

In this section we shall give the first application of AU-spaces by proving a uniqueness theorem for convolution operators that generalizes the uniqueness theorem for the heat equation.[4] The problem consists in the following. Suppose we know that $f(x,t) \in C^{\infty}(\mathbb{R}^{n+1})$ satisfies the equation

$$(20) \quad L^*f(x,t) = D_t^q f(x,t) + \mu_1 * D_t^{q-1} f(x,t) + \dots + \mu_q * f(x,t) = 0 ,$$

where $\mu_1, \dots, \mu_q$ are given distributions with compact support acting on the x-variables, i.e. $\mu_j \in \mathcal{E}'(\mathbb{R}^n)$. Furthermore, assume that f has zero Cauchy data, i.e. $D_t^j f(x,0) \equiv 0$ for $j = 0,1,\dots,q-1$. When can we conclude that $f \equiv 0$?

It is well known that even for differential operators this does not hold unless the hyperplane $\{(x,t): t=0\}$ is non-characteristic [27]. For the characteristic case one has to impose additional restrictions upon f, e.g. certain growth conditions on f (cf. [48]). Here we shall impose growth conditions on the x-variables only.

Let us set

$$\tilde{\Phi}(x,t) = \Phi(x) + \Phi_0(t) ,$$

where $\Phi_0(t) = 0$ for $|t| < 1$ and $\Phi_0(t) = +\infty$ for $|t| \geq 1$. We shall assume that the function f is in $\mathcal{E}(\tilde{\Phi})$. Actually, we can go even further and study the case when f is not a solution to (20), but satisfies this equation only approximately, i.e. when $f \in \mathcal{E}_B(L;\tilde{\Phi})$ and $D^{\alpha} f(x,0) \equiv 0$ for all multiindices $\alpha = (\alpha_1, \dots, \alpha_{n+1})$. Theorem 2 gives conditions on B and $\tilde{\Phi}$ which imply $f \equiv 0$. Theorem 4 represents an analogous result for an "overdetermined" system.

Remark 2. In the case when f actually solves equation (20), the condition,

(21) $$D^{\alpha}f(x,0) \equiv 0 \quad \text{for all } \alpha ,$$

is a consequence of $D_t^j f(x,0) \equiv 0$ for $1 \leq j \leq q-1$. If f satisfies condition (20), f is said to have zero Cauchy data. The variables dual to $(x,t) \in \mathbb{R}^{n+1}$ will be denoted by $w = (z,s) \in \mathbb{C}^{n+1}$. Finally, in this section a convex p-sequence B is always assumed to be of the form $B = \{b_j\}$, $b_j = b_{j_1}^{(1)} \ldots b_{j_p}^{(p)}$ where all the sequences $B^{(k)} = \{b_j^{(k)}\}$ are convex (cf. §1).

We shall also need some results from the theory of Denjoy-Carleman classes [38,42].

Definition 2. Let M be a fixed convex sequence. Then the Denjoy-Carleman class $\mathcal{E}_M$ is defined as the space of all functions $f \in C^{\infty}([0,1])$ such that

$$|f^{(j)}(x)| \leq c_0 c_1^j b_j \quad \text{for all } j \geq 0 \text{ and } x \in [0,1]$$

with some constants c_0, c_1 depending on f.

A class $\mathcal{J}$ of C^{∞}-functions (on [0,1]) is called quasi-analytic, if no function $f \not\equiv 0$ in $\mathcal{J}$ can vanish together with all its derivatives at any point. If $\mathcal{E}_M$ is a quasi-analytic class, the sequence M will be called quasi-analytic.

Denjoy-Carleman Theorem. The following three conditions are equivalent:

(a) $\mathcal{E}_M$ is quasi-analytic ;

(b) $\int_1^{\infty} \log(\lambda_M(u))u^{-2}\,du = \infty$ (cf. (3)) ;

(c) $\sum_{j\geq 1} b_j^{-1/j} = \infty$.

One way of generating quasi-analytic sequences is the following. Let $R(u)$ be a positive strictly increasing function of $u \geq 0$ such that $\log (R(u))$ is a convex function of $\log u$ and satisfying for all $j \geq 0$,

$$\lim_{u\to\infty} \frac{u^j}{R(u)} = 0 ;$$

then, we define a sequence $M = \{m_j\}$ by

$$m_j = \max_{u\geq 0} \frac{u^j}{R(u)} \tag{22a}$$

Then, as it is shown in [38], one can find positive constants α,β such that

$$\log (R(u)) - \log (1+u) - \beta \leq \log (\lambda_M(u)) \leq \log (R(2u)) + \alpha. \tag{22b}$$

Therefore it follows that the class $\mathcal{E}_M$ is quasi-analytic if and only if

$$\int_1^\infty \log (R(u))u^{-2}\, du = \infty .$$

Given a convex sequence B and a positive integer q we can define a new sequence $\{m_j\}$, denoted by B/q, by setting first $R(u) = \lambda_B(u^q)$ for $u \geq 0$; and, defining m_j as in (22). Then, we will have

$$b_{[\frac{j}{q}]} \leq m_j \leq b_{[\frac{j}{q}]+1}$$

where $[j/q]$ denotes as usual the integral part of j/q.

The following modification $C = \{c_j\}$ of a given quasi-analyt sequence B will be also used in the sequel (cf. (33) below). If $B = \{b_j\}$, let $C = \{c_j\}$ be defined by

$$c_j = \max (b_j, j!) \qquad (j = 0,1,\ldots) .$$

It is clear that C is a convex sequence and

$$\lambda_C(u) \leq e^{|u|} ; \quad \lambda_C(u) \leq \lambda_B(u) .$$

However we claim that C is also quasi-analytic. Since B is quasi-analytic we may assume $j! \geq b_j$ for infinitely many j; otherwise the result would be clear. Then, there is a sequence of integers j_k such that $j_o = 1$, $2j_k \leq j_{k+1}$ and $b_{j_k} \leq j_k!$. Since C is a convex sequence, $c_j^{1/j}$ must be increasing; hence

$$\sum_{j=j_k+1}^{j_{k+1}} c_j^{-1/j} \geq (j_{k+1}-j_k) c_{j_{k+1}}^{-1/(j_k+1)}$$

$$\geq (j_{k+1}-j_k)(j_k!)^{-1/j_{k+1}} \geq \frac{1}{2} j_{k+1}(j_{k+1})^{-1} = \frac{1}{2} .$$

Thus $\sum c_j^{-1/j} = \infty$ and by the Denjoy-Carleman theorem, the sequence C is quasi-analytic.

If $\mathcal{E}_M$, $\mathcal{E}_N$ are two quasi-analytic classes, their "sum" $\mathcal{E}_M + \mathcal{E}_N = \{f+g\colon f \in \mathcal{E}_M, g \in \mathcal{E}_N\}$ is not necessarily quasi-analytic [1,38]. Nevertheless, the convex regularization yields a partial result in this direction:

Lemma 1. If $M = \{j!\}$ and N is quasi-analytic, then $\mathcal{E}_M + \mathcal{E}_N$ is also quasi-analytic.

Let us first sketch the intuitive idea which underlies both the statement and the proof of Theorem 1. Our objective will be to find functions H(y,w), analytic in w and belonging to a fixed quasi-analytic class on the interval $0 \leq y \leq 1$; and, moreover, such that

(i) The functions of the form H(1,w) form a total set in $\hat{\mathcal{E}}_B'(L;\tilde{\Phi})$.

(ii) For all $j \geq 0$,

$$\operatorname{supp} \frac{\partial^j H(0,w)}{\partial y^j} \subset \{w \in \mathbb{C}^{n+1}\colon s = 0\} .$$

If $f \in \mathcal{E}_B(L;\hat{\Phi})$, then by Theorem 1 and Remark 6, I there exists a majorant k (in the AU-structure described in Th.1) and a Radon measure $d\nu(w)$ such that

$$f(x,t) = \int_{\mathbb{C}^{n+1}} e^{i<(x,t),w>} \frac{d\nu(w)}{k(w)} .$$

(iii) Furthermore, it will be shown below that the functions

$$h(y) \overset{\text{def}}{=} <f(\cdot),H(y,\cdot)> = \int_{\mathbb{C}^{n+1}} H(y,w) \frac{d\nu(w)}{k(w)}$$

are in a fixed quasi-analytic class. Now if f has zero Cauchy data, then by (ii), $h^{(j)}(0) = 0$ for $j = 0,1,\ldots$. Then (iii) implies $h(1) = 0$ for all H, and by (i) we obtain $f = 0$.

First we need the following lemma.

Lemma 2. Let $\zeta = \sigma + i\tau \in \mathbb{C}$ denote the complex variable and $\psi(\tau)$ an even positive convex function for which

$$(23) \qquad |\tau|^a = \mathcal{O}(\psi(\tau)) \quad \text{with some } a > 1 .$$

Then there are non-zero entire functions F such that

$$(24) \qquad |F(\zeta)| = \mathcal{O}(\exp(-c|\sigma|^a + \Psi(c'\tau))$$

where c, c' are some positive constants depending on F. Moreover, for any such function F, the set of linear combinations of functions of the form $e^{i\alpha\zeta}F(\zeta+\beta)$ $(\alpha,\beta$ real) is dense in $\hat{\mathcal{E}}'(\Psi^*)$; here Ψ^* denotes the Young conjugate of the function Ψ. Since any of the functions $e^{i\alpha\zeta}F(\zeta+\beta)$ satisfies (24), the set of all functions satisfying (24) is also dense in $\hat{\mathcal{E}}'(\Psi^*)$.

Proof: The existence of entire functions $F \neq 0$ satisfying (24) is shown, for instance, in [33]. Assume that F is such a function and set

$$\hat{f}(\zeta) = F(-\zeta) .$$

Then $f \in \mathcal{E}'(\Psi^*) \cap \mathcal{S}$ where $\mathcal{S}$ is the Schwartz space of rapidly decreasing functions [46]. We have to prove that the set $\mathcal{M}$ of all linear combinations of functions of the form $e^{i\alpha x} f(\beta - x)$ (α,β real) is dense in $\mathcal{E}'(\Psi^*)$. Let us first reduce this problem to showing that every function of the form $g(x)f(\beta-x)$ (with $g \in C_0^\infty$ and β real) is in the closure of $\mathcal{M}$. One can easily construct a sequence of functions g_m, $g_m \in C_0^\infty$, such that for every $h \in \mathcal{E}(\Psi^*)$, $g_m h \to h$ in $\mathcal{E}(\Psi^*)$. Then, if h is orthogonal to $\mathcal{M}$, i.e. $\langle T,h\rangle = 0$ for all $T \in \mathcal{M}$, we shall have, for all $m \geq 1$ and $\beta \in \mathbb{R}$,

(25) $$\langle g_m(x)f(\beta-x),h(x)\rangle = \langle f(\beta-x),g_m(x)h(x)\rangle = 0 .$$

Since $g_m h \in C_0^\infty$ and $f \in \mathcal{S}$, equality (25) says that

(26) $$f*(g_m h)(\beta) = 0 \qquad \text{for all } \beta \in \mathbb{R} \text{ and } m \geq 1.$$

Applying the Fourier transform to (26) we obtain

$$F(-\zeta)\widehat{g_m h}(\zeta) = 0 \qquad \text{for all } \zeta \in \mathbb{R}.$$

Since F is a non-zero analytic function, we obtain from here that $\widehat{g_m h} = 0$, for all m, i.e. $h = 0$. This shows that it suffices to prove that all functions $g(x)f(\beta-x)$, $\beta \in \mathbb{R}$, $g \in C_0^\infty$, are in the closure of $\mathcal{M}$.

Let $\mathcal{O}_M$ be the Schwartz space of all C^∞-functions of polynomial growth in $\mathbb{R}$ [46]. The topology of $\mathcal{O}_M$ has the following property. Let ρ be a positive continuous function satisfying for all $m = 0,1,\ldots$,

$$\lim_{|x|\to\infty} \frac{\rho(x)}{|x|^m} = 0 ;$$

and, let $\{h_\gamma\}_\gamma$ be a net in $\mathcal{O}_M$ such that $h_\gamma \to 0$. Then

(27) $$\sup_{-\infty<x<\infty} \{\rho(x)|h_\gamma^{(k)}(x)|\} \to 0$$

for all $k = 0,1,\ldots$. We claim that for each fixed $T \in \mathcal{E}'(\Psi^*)$ the mapping

(28) $$h \mapsto hT$$

of $\mathcal{O}_M$ into $\mathcal{E}'(\Psi^*)$ is continuous. Indeed, given $h_\gamma \to 0$ in $\mathcal{O}_M$ and a bounded set $\{g\}$ in $\mathcal{E}(\Psi^*)$, we have for every $\varepsilon > 0$ and $k \geq 0$,

$$\sup_x \{|(h_\gamma g)^{(k)}(x)| \exp(-\Psi^*(\varepsilon x))\} \leq C_k \sup_{x;1\leq j\leq k} \{|g^{(j)}(x)|\exp(-\Psi^*(\tfrac{\varepsilon x}{2}))\}$$

$$\times \sup_{x;1\leq j\leq k} \{|h_\gamma^{(j)}(x)|\rho(x)\}$$

for some positive constant C_k and

$$\rho(x) = \exp[\Psi^*(\tfrac{\varepsilon x}{2}) - \Psi^*(\varepsilon x)] .$$

Therefore $h_\gamma g \to 0$ in $\mathcal{E}(\Psi^*)$ uniformly with respect to $\{g\}$, and the continuity of the mapping (28) follows.

Every function $g \in C_0^\infty$ can be approximated in $\mathcal{O}_M$ by linear combinations of $e^{i\alpha x}$, $\alpha \in \mathbb{R}$, for it suffices to consider the Riemann sums of the integral

$$g(x) = \frac{1}{2\pi} \int_{-\infty}^{\infty} e^{-ixv} \hat{g}(v)\, dv .$$

Hence, for any β real, the function $g(x)f(\beta - x)$ is in the closure of $\mathcal{M}$ in $\mathcal{E}'(\Psi^*)$. This completes the proof of Lemma 1.

Since each μ_j in (20) is a distribution with compact support, there exist positive constants A, B and C such that, for all $j = 1,2,\ldots,g$, we have

$$|\hat{\mu}_j(z)| \leq C(1+|\mathrm{Re}\, z|)^B e^{A|\mathrm{Im}\, z|} \leq \frac{C}{2}(1+|\mathrm{Re}\, z|)^{2B} + \frac{C}{2} e^{2A|\mathrm{Im}\, z|}$$

More generally, we can assume that we are given a function $\rho(u)$, $u \geq 0$, which is positive, continuous, strictly increasing and such that for any $\delta > 0$ there are positive numbers δ' and δ'' satisfying

(29) $$\delta\rho(u) \leq \rho(\delta' u) + \delta''$$

(examples of ρ: $\rho(u) = e^u$; $\rho(u) = u^m$, etc.). Moreover, we shall assume that the Fourier transforms of the distributions μ_j satisfy the estimates

$$(30) \qquad |\hat{\mu}_j(z)|^{1/j} \leq C_1 + C_2|\text{Re } z|^D + \rho(|\text{Im } z|)$$

where C_1, C_2 and D are some non-negative constants. Let us observe that $\rho(u) \nearrow \infty$ (except when all μ_j are zero, and in this case the answer is well known [23]). Therefore the inverse function $\rho_{-1}(u)$ of ρ is well defined. The function $\Psi(u)$ will be assumed convex, positive and such that the function

$$(31) \qquad R(u) = \exp\ (\Psi(\rho_{-1}(u)))$$

satisfies the above conditions on R (cf. the text following the Denjoy-Carleman theorem). Then by (22a,b) there exists a convex sequence $M = \{m_j\}$ such that for some positive constant C_3,

$$(32) \qquad \lambda_M(\tfrac{u}{2}) \leq C_3\ R(u)\ ;$$

and, we may further assume (cf. the modification C of B preceding Lemma 1) that

$$(33) \qquad \lambda_M(u) \leq e^{|u|}\ .$$

Let us recall that M is quasi-analytic if and only if

$$\int_1^\infty \Psi(\rho_{-1}(u))u^{-2}\ du = \infty.$$

We can finally state

<u>Theorem 2</u>. Assume that Ψ is as above and satisfies (23) with an $a \geq D$. Furthermore, suppose that

$$(34) \qquad \mathcal{E}_M + \mathcal{E}_{B/q} \text{ is a quasi-analytic class.}$$

Let $\Phi(x) = \Psi^*(|x|)$ and $\tilde{\Phi}(x,t)$ be the function defined in the beginning of this section. Then

$$\{f \in \mathcal{E}_B(L;\tilde{\Phi}) \quad \text{and} \quad D^\alpha f(x,0) = 0 \ (\forall\alpha)\} \Rightarrow f = 0 .$$

Proof: Lemma 2 and the definition of $\tilde{\Phi}$ imply that the set of functions of the form

$$e^{i\sigma s} F_1(z_1)\ldots F_n(z_n) \tag{35}$$

where σ is real and F_i satisfy condition (24), is total in $\hat{\mathcal{E}}'(\tilde{\Phi})$. In particular, this set is total in $\hat{\mathcal{E}}_B'(L;\tilde{\Phi})$.

Let us define the family $\mathcal{H}$ of functions $H(y,w)$, $w \in \mathbb{C}^{n+1}$, $0 \leq y \leq 1$, by

$$\mathcal{H} = \{H(y,w) = e^{iy\sigma s} F_1(z_1)\ldots F_n(z_n),\ F_i \text{ as in } (35)\} .$$

Then

(36) the set $\{H(1,w) : H \in \mathcal{H}\}$ is total in $\hat{\mathcal{E}}_B'(L;\tilde{\Phi})$.

Moreover, for $H \in \mathcal{H}$ and $j = 0,1,\ldots$,

$$\frac{\partial^j}{\partial y^j} H(y,w) = (i\sigma s)^j H(y,w) ;$$

therefore,

$$\frac{\partial^j}{\partial y^j} H(0,w) = \text{Fourier transform of } \delta_t^{(j)} \otimes T_x,\ T_x \in \mathcal{E}_x'(\Phi) ,$$

i.e.,

(37) $\frac{\partial^j}{\partial y^j} H(0,w)$ acts only on the Cauchy data of the functions in $\mathcal{E}_B(L;\tilde{\Phi})$.

Using the Fourier representation of $f(x,t)$, it follows from the estimates below that for every $H \in \mathcal{H}$ the function $h(y)$ defined in (iii) above is a C^∞-function and

$$h^{(j)}(y) = \int \frac{\partial^j H}{\partial y^j}(y,w) \frac{d\nu(w)}{k(w)} \qquad (j \geq 0) . \tag{38}$$

Let us set

$$A = \{w = (z,s) : |s| \leq \max_{1 \leq j \leq q} |2q\hat{\mu}_j(z)|^{1/j}\} .$$

Then

$$B = \mathbb{C}^{n+1} \setminus A \subseteq \{w : |\hat{L}(w)| \geq \frac{1}{2} |s|^q\} .$$

Given $\varepsilon > 0$, by (29) we can choose $\varepsilon', \varepsilon'' > 0$ such that, for all $u \geq 0$,

$$(39) \qquad 8q\varepsilon\rho(u) \leq \rho(\varepsilon' u) + 4\varepsilon'' .$$

If $w \in A$ and $|\text{Im } z|$ are so large that

$$4q\varepsilon\rho(|\text{Im } z|) - 2\varepsilon'' > 0 ,$$

then, using (30), (32), (33), (39) and the convexity of λ_M, we obtain

$$\begin{aligned}
\lambda_M(\varepsilon s) &= \lambda_M(|\varepsilon s|) \\
&\leq \lambda_M(2q\varepsilon C_1 + 2q\varepsilon C_2 |\text{Re } z|^D + 2q\varepsilon\rho(|\text{Im } z| + \varepsilon'' - \varepsilon'')) \\
&\leq \frac{1}{2} \lambda_M(4q\varepsilon C_1 + 4q\varepsilon C_2 |\text{Re } z|^D + 2\varepsilon'') + \frac{1}{2} \lambda_M(4q\varepsilon\rho(|\text{Im } z|) - 2\varepsilon'') \\
&\leq C_4 \exp(4q\varepsilon C_2 |\text{Re } z|^D) + \frac{1}{2} C_3 R(8q\varepsilon\rho(|\text{Im } z|) - 4\varepsilon'') \\
&\leq C_4 \exp(4q\varepsilon C_2 |\text{Re } z|^D) + \frac{1}{2} C_3 \exp(\Psi(\varepsilon' |\text{Im } z|)) .
\end{aligned}$$

For the remaining points w in A we have

$$\lambda_M(\varepsilon s) = C_5 \exp(2q\varepsilon C_2 |\text{Re } z|^D) .$$

On the other hand, for $y \in [0,1]$, $j = 0,1,\ldots,$ and $H \in \mathcal{H}$, we get

$$\left|\frac{\partial^j}{\partial y^j} H(y,w)\right| \leq c''|\sigma|^j |s|^j \exp[|\sigma \text{ Im } s| - c|\text{Re } z|^a + \Psi(c'|\text{Im } z|)]$$

for some positive c, c', c'' and σ real. Set $C_6 = \max\{C_4, C_5\}$ and $\varepsilon > 0$ so small that $4q\varepsilon C_2 \leq c$. Then

$$\left|\frac{\partial^j}{\partial y^j} H(y,w)\right| \leq c''\left(\frac{|\sigma|}{\varepsilon}\right)^j m_j \lambda_M(\varepsilon s) \exp\left\{|\sigma \text{ Im } s| - c|\text{Re } z|^a + \Psi(c'|\text{Im } z|)\right\}$$

$$\leq C_6 c''\left(\frac{|\sigma|}{\varepsilon}\right)^j m_j \exp\left\{|\sigma \text{ Im } s|+4q\varepsilon C_2|\text{Re } z|^D - c|\text{Re } z|^a + \Psi(c'|\text{Im } z|)\right\}$$

$$+ \frac{C_3}{2} c''\left(\frac{|\sigma|}{\varepsilon}\right)^j m_j \exp\left\{|\sigma \text{ Im } s| + \Psi(c'|\text{Im } z|) + \Psi(\varepsilon'|\text{Im } z|)\right\}$$

$$\leq C_7\left(\frac{|\sigma|}{\varepsilon}\right)^j m_j\ k(w)\ .$$

In the proof of the last inequality we have used the inequality $D \leq a$ and the fact that the function k dominates all the other factors. The last estimates show that

$$\int_A H(y,w)\ \frac{d\nu(w)}{k(w)} \in \mathcal{E}_M\ . \tag{40}$$

In the set B we have $|\hat{L}(w)| \geq \frac{1}{2}|s|^q$; and, since k contains a factor larger than $\lambda_B(2|\hat{L}(w)|) \geq \lambda_B(|s|^q)$,

$$\int_B H(y,w)\ \frac{d\nu(w)}{k(w)} \in \mathcal{E}_{B/q}\ . \tag{41}$$

Relations (40) and (41) show that $h \in \mathcal{E}_M + \mathcal{E}_{B/q}$. Since $D^\alpha f(x,0) \equiv 0$ for all α, we get from (37) and (38) that $h^{(j)}(0) = 0$ for all j. The quasi-analyticity of the class $\mathcal{E}_M + \mathcal{E}_{B/q}$ then implies $h(1) = 0$, i.e.

$$< H(1,w), f> = 0 \qquad \text{for all } H \in \mathcal{H}\ ,$$

which, as we showed in (36), implies $f \equiv 0$. The theorem is completely proved.

Instead of inequality (30) we can use the inequality

$$|\hat{\mu}_j(z)|^{1/j} \leq C_o + \sum_{i=1}^n C_i|\text{Re } z_i|^{D_i} + \sum_{i=1}^n \rho_i(|\text{Im } z_i|) \tag{42}$$

with C_o and C_i positive, $D_i \geq 0$ and ρ_i sastisfying the same conditions as ρ; and,with Ψ_i being even, convex and positive functions satisfying (23) for $a_i \geq D_i$. The function R(u) is chosen so that R is dominated by all exponentials $\exp(\Psi_i((\rho_i)_{-1}(u)))$, and satisfies the same conditions as in Theorem 2. Then, for the corresponding M (cf. (22a)), the same proof yields the following

Theorem 3. Assume that (34) holds and set $\Phi(x) = \sum_{i=1}^{n} \Psi_i^*(x_i)$. Then the conclusion of Theorem 2 holds.

Similarly one can study 'overdetermined' systems of the form

$$L_1*f(x,t) = D_t^{q_r} f(x,t) + \mu_{r,1}*D_t^{q_r-1} f(x,t) + \dots + \mu_{r,q_r}*f(x,t)$$

$$(r=1,\dots,p)$$

where the distributions $\mu_{r,j}$ act only on the x-variable. One has to introduce a new convex sequence C derived from the p-sequence B by means of

$$S(u) = \lambda_B(u^{q_1},\dots,u^{q_p})$$

as in (22a). Let us observe that C is a quasi-analytic sequence whenever one of the convex sequences $B^{(r)}/q_r$ is quasi-analytic; indeed,

$$S(u) \geq \lambda_{B^{(r)}}(u^{q_r}) \qquad \text{for } r = 1,\dots,p \; .$$

Moreover, if (42) holds for all $\mu_{r,j}$, we can define M and Φ as in Theorem 3 and obtain

Theorem 4. Let Φ, M and C be as above. Assume that (34) holds for the class $\mathcal{E}_M + \mathcal{E}_C$. Then any function f, $f \in \mathcal{E}_B(L_1,\dots,L_p;\tilde{\Phi})$, such that $D^\alpha f(x,0) \equiv 0$ for all α, is identically zero.

Example 1. Let

$$L*f(x,t) = D_t f(x,t) + \mu*f(x,t)$$

and

$$\Phi(x) = |x| \log(1+|x|) \; .$$

Then for any μ with compact support and any quasi-analytic sequence B, the hypotheses of Theorem 2 are satisfied. In fact, let $\rho(u) = \Psi(u) = e^{A|u|}$ for some $A > 0$. Then $\lambda_M(u) = e^{|u|}$ and $M = \{j!\}$. By Lemma 1, the class $\mathcal{E}_M + \mathcal{E}_B$ is quasi-analytic.

Example 2. Let us consider the heat equation

$$L * f(x,t) = \frac{\partial}{\partial t} f(x,t) - \frac{\partial^2}{\partial x^2} f(x,t) \qquad (x,t) \in \mathbb{R}^2 .$$

then for $\Phi(x) = |x|^2$ and any quasi-analytic sequence B we can repeat the method of Example 1, namely, set $\rho(u) = \Psi(u) = u^2$, etc.

Example 3. A slight modification of the preceding operator is the difference-differential operator

$$L * f(x_1,x_2,t) = D_t f(x_1,x_2,t) - D^2_{x_1} f(x_1,x_2+1,t)$$

which can also be studied with the aid of Theorem 3. (Namely, we set $\Phi_1(x_1) = |x_1|^2$, $\Phi_2(x_2) = |x_2| \log(1+|x_2|)$ and take for B an arbitrary quasi-analytic sequence, etc.).

CHAPTER IV

The Fundamental Principle

§1. FORMULATION OF THE THEOREM AND AUXILIARY LEMMAS

In this chapter we shall prove the main result on AU-spaces. The motivation for the theorem can most easily be seen in the case $n = 1$. In fact, this case dates back to Leonhard Euler.

Let T be a distribution solution of a homogeneous linear differential equation with constant coefficients,

$$(1) \qquad a_m \frac{d^m T}{dt^m} + a_{m-1} \frac{d^{m-1} T}{dt^{m-1}} + \dots + a_o T = 0 .$$

Then T is a C^∞-function on the real line [46]. Moreover, T is an exponential polynomial, i.e.

$$(2) \qquad T(x) = \sum_{\ell=1}^{r} \sum_{j=0}^{J_\ell - 1} c_j \, x^j \, e^{i\alpha_\ell x} ,$$

where α_ℓ are the roots of the polynomial

$$(3) \qquad P(z) = a_m (iz)^m + a_{m-1} (iz)^{m-1} + \dots + a_o ;$$

J_ℓ is the multiplicity of α_ℓ, $J_1 + \dots + J_r = m$; and, the c_j's are constants depending on T.

On the other hand, since we know that the space $\mathcal{D}'(\mathbb{R})$ is an AU-space (cf. Th. 1, II), T must have a Fourier representation of the form

$$(4) \qquad T(x) = \int_{\mathbb{C}} e^{ixz} \frac{d\mu(z)}{k(z)} ;$$

(Corollary 2, I and Remark 6, I). If all the roots of P are simple, relation (2) can be viewed as a Fourier representation of this kind, but this particular representation has an additional property, namely, the measure μ occurring in (2) is such that

$$\text{(5)} \qquad \operatorname{supp} \mu \subseteq V_P = \{z : P(z) = 0\} = \{\alpha_1, \ldots, \alpha_m\} .$$

Conversely, if (5) holds, then $d\mu(z)/k(z)$ is a linear combination of the Dirac measures $\delta_{\alpha_1}, \ldots, \delta_{\alpha_m}$; hence, representation (4) reduces to

$$T(x) = \sum_{j=1}^{r} c_j \, e^{i\alpha_j x} ,$$

and T obviously solves the equation (1). However, if the roots of P are not simple, the Fourier integrals (4) satisfying condition (5) no longer furnish all the solutions of the equation. In this case we may proceed as follows. Since

$$x^j \, e^{ixz} = i^{-j} \frac{d^j}{dz^j} (e^{ixz}) ,$$

it is natural to introduce at every point α_ℓ (i.e., at every irreducible component of the algebraic variety V_P) the differential operators

$$\text{(6)} \qquad \partial_{j,\ell} = i^{-j} \frac{d^j}{dz^j} ; \qquad j = 0,1,\ldots,J_\ell .$$

Then we can write equality (2) in the form

$$\text{(7)} \qquad T(x) = \sum_{\ell=1}^{r} \sum_{j=0}^{J_\ell - 1} \int \partial_{j,\ell} \, e^{ixz} \frac{d\mu_{j,\ell}(z)}{k(z)} ,$$

with $\operatorname{supp} \mu_{j,\ell} \subseteq \{z = \alpha_\ell\}$. Moreover, it is clear that now formula (7) yields all the solutions of equation (1).

Next we can ask what can be said in the case $n > 1$; or, more exactly, whether one still has a similar description of the "general" solution of a homogeneous linear partial differential equation with constant coefficients. However natural this question may appear, the reply has always been that it makes little sense to look for general solutions when $n > 1$. Rather one always looked for particular solutions satisfying additional conditions (e.g. boundary or initial conditions, etc); and, in order to determine the appropriate conditions, it was necessary to classify the PDE's in the well known fashion. Thus, Euler's approach (cf. (2)) has always been considered as limited to ordinary differential equations. Nevertheless, it recently turned out that Euler's method may indeed be generalized, though in a very sophisticated manner, to partial differential equations with constant coefficients, and also to systems of such equations . Moreover, this approach became a powerful tool for investigating different properties of such PDE's . Very roughly, this is the essence of the theorem discovered in 1960 by L. Ehrenpreis [18] and called by him the fundamental principle. The next section is devoted to the proof of this result. For the sake of simplicity we shall limit ourselves to the case of one partial differential equation. The main corollary of the fundamental principle reads as follows:

Let $P(z)$ be a polynomial in $z = (z_1,\ldots,z_n) \in \mathbb{C}^n$; D as in Chap. I; and W a suitable AU-space of distributions. Then each $T \in W$ satisfying equation

$$P(D)T = 0 \ , \tag{8}$$

can be represented in the form

$$T(x) = \sum_{j=1}^{p} \int_{V_j} \partial_j e^{ixz} \frac{d\mu_j(z)}{k(z)} \ , \tag{9}$$

where the V_j's are $\mathcal{X}$-subvarieties (cf. below) of the algebraic variety $V_P = \{z: P(z) = 0\}$ (V_j's are not necessarily all different), and the ∂_j's are certain differential operators associated with equation (8). (The class of spaces W for which this statement is proved below is the class of PLAU-spaces, cf. Def. 3, I.) Representation (9) is obviously the desired generalization of the Euler formula (2) to the case $n > 1$. However, it is not difficult to briefly describe the main theorem itself.

Let W be an AU-space and T an element of W satisfying (8). Then

$$\langle P(D)T, f\rangle = \langle T, P(D)^t f\rangle = \langle \hat{T}, P(z)\hat{f}(z)\rangle = 0 ,$$

for all $f \in U$. Hence $\hat{T}$ is a continuous linear functional on the space $\hat{U}/P\hat{U}$; and, conversely, every continuous linear functional on $\hat{U}/P\hat{U}$ defines an element T of W satisfying equation (8). If the functions $F_1, F_2 \in \hat{U}$ belong to the same coset modulo $P\hat{U}$, their restrictions to the set V_P must coincide. If $\tilde{F} = F_1|V_P = F_2|V_P$,* then the function $\tilde{F}$ is continuous on V_P and satisfies the same growth conditions as the elements of $\hat{U}$. Moreover, we also know that the function $F_1 - F_2$ vanishes at every point of V_P with (at least) the same order as the polynomial P. This hints to the possibility of dividing the variety V_P into a finite number of parts V_j for which there exist differential operators ∂_j on V_j such that not only $(F_1-F_2)|V_P = 0$, but also

(10) $$(\partial_j(F_1-F_2))|V_j = 0 .$$

The theorem we are going to prove states that the converse is also true, i.e. if $F_1, F_2 \in \hat{U}$ satisfy (10), then $F_1 \equiv F_2 \pmod{P\hat{U}}$; moreover, given any analytic function $\tilde{F}$ on V_P (i.e. a restriction to V_P of an analytic function) satisfying on V_P the same growth conditions as restrictions to V_P of functions in $\hat{U}$, then there exists a function F in $\hat{U}$ such that for all j,

$$\partial_j F|_{V_j} = \partial_j \tilde{F}|_{V_j} .$$

Let us call the set $\mathcal{V} = (V_j, \partial_j)_j$ a <u>multiplicity variety</u>. A system of functions $\{\tilde{F}_j = \partial_j \tilde{F}\}_j$, where $\tilde{F}$ is an analytic function on $\mathcal{V}$, is called an analytic function on $\mathcal{V}$. The vector space $\hat{U}(\mathcal{V})$ is defined as the space of all analytic functions on $\mathcal{V}$ bounded on $\bigcup_j V_j$ by functions k, $k \in \mathcal{K}$, where $\mathcal{K}$ is the AU-structure of W. These bounds obviously define a natural l.c. topology on $\hat{U}(\mathcal{V})$. The above

*$F|V_P$ = restriction of F to V_P.

mentioned theorem can then be formulated as follows:

Fundamental principle.

Let W be a PLAU-space and P a polynomial. Then there is a multiplicity variety $\mathcal{V}$ defined by means of the algebraic variety $V_P = \{z: P(z) = 0\}$ such that the mapping

$$\hat{U}/P\hat{U} \to \hat{U}(\mathcal{V}) \tag{11}$$

is a topological isomorphism.

As can be easily seen, this theorem implies the Fourier representation for all solutions of (8) (cf. Corollary B below).

Remarks: 1. The fundamental principle holds for systems of partial differential equations with constant coefficients (cf. [23], Chap. IV). However then the definition of the multiplicity variety as well as the whole proof becomes more complicated, although everything proceeds along similar lines as in the case of one equation [23,41].

2. The idea of the proof consists in extending each function $F \in \hat{U}(\mathcal{V})$ locally from V to the surrounding space, and then, correcting these extensions so that they define a function in $\hat{U}$ (in particular, special care must be taken of the bounds). In other words, one must show the vanishing of a certain cohomology group. Actually, the proof closely follows the Cartan-Oka-Serre proof of the vanishing of the cohomology groups $H^i(\mathbb{C}^n, \mathcal{F})$, $i > 0$, where $\mathcal{F}$ is a coherent analytic sheaf [25]. However, knowledge of the latter proof will not be presupposed in the sequel.

Now we are ready to start with definitions and some auxiliary facts which will be needed in the proof.

Definition 1. If w is a point of $\mathbb{C}^{n+1}$, we shall write $w = (s,z)$ where

$s \in ¢$ and $z \in ¢^n$. An analytic function $P(w)$, defined in an open set $¢ \times B$, $B \subseteq ¢^n$, will be called a <u>distinguished polynomial</u> in s of degree m, if we can write

$$P(s,z) = s^m + P_1(z)s^{m+1} + \ldots + P_m(z) ,$$

with the coefficients $P_j(z)$ analytic in B.

<u>Remark</u>. 3. Let us observe that by Definition 1, a <u>distinguished polynomial</u> $P(w)$ <u>in</u> s <u>is not necessarily a polynomial in</u> w. On the other hand, by an appropriate nonsingular change of variables any polynomial $Q(\zeta)$ in $¢^{n+1}$ becomes a nonzero multiple of a distinguished polynomial. Namely, let

$$Q(\zeta) = \sum_{j=0}^{m} Q_j(\zeta) ,$$

where $Q_j(\zeta)$ is a homogeneous polynomial of degree j. Let $Q_m(a) \neq 0$ for some $a \in ¢^{n+1}$; in particular, $a \neq 0$. Next we choose <u>n</u> arbitrary points $b^{(j)} \in ¢^{n+1}$ such that the (n+1) vectors $a_1, b^{(1)}, b^{(2)}, \ldots, b^{(n)}$ are linearly independent. Then there are points $s \in ¢$, $z \in ¢^n$ satisfying the system of equations

$$\zeta_i = a_i s + \sum_{j=0}^{n} b_i^{(j)} z_j .$$

The change of coordinates $\zeta \mapsto w = (s,z)$ is clearly nonsingular and the polynomial $P(w)$ defined by $P(w) = Q(\zeta)$ has the same degree as Q. Moreover, if $P_m(w)$ denotes the homogeneous part of degree <u>m</u> in P, then $P_m(w) = Q_m(\zeta)$, and

$$P_m(s,0) = Q_m(as) = s^m Q_m(a) .$$

This shows that

$$P(s,z) = s^m Q_m(a) + \text{terms of degree} \leq m-1 \text{ in } s \quad (Q_m(a) \neq 0),$$

which proves the above assertion.

The following lemma is known as the Weierstrass division theorem [25]; however, the vanishing of the coefficients $P_j(z)$ at the origin will not be assumed here. Later we shall discuss a local version of this theorem.

Lemma 1. Let $F(s,z)$ be analytic in the open set $\mathbb{C} \times B$, $B \subseteq \mathbb{C}^n$, and let $P(s,z)$ be a distinguished polynomial in s of degree m in $\mathbb{C} \times B$. Then

$$F(s,z) = G(s,z)P(s,z) + R(s,z)$$

where $G(s,z)$ is analytic in $\mathbb{C} \times B$, and

$$R(s,z) = \sum_{j=0}^{m-1} s^j R_j(z)$$

with the coefficients R_j analytic in B. Moreover, let $C \overset{\text{def}}{=} C(z) = \max\{|P_j(z)|: j = 1,\dots,m\}$. Then

$$|R_j(z)| \leq 2(1+C) \max\{|F(\sigma,z)|: |\sigma| \leq 2(1+C)\};$$

and, if $r = \max\{|s|,1\}$, then

$$|G(s,z)| \leq (r+C)^{-m} \max\{|F(\sigma,z)|: |\sigma| \leq 2(r+C)\}.$$

Proof: First, let us observe that by the definition of $P(s,z)$,

$$s^m = -(P_1(z)s^{m-1} + \dots + P_m(z)) + P(s,z);$$

hence, we obtain

$$s^{m+1} = [P_1^2(z) - P_2(z)]s^{m-1} + [P_1(z)P_2(z) - P_3(z)]s^{m-2} + \dots$$

$$+ P_1(z)P_m(z) + P(s,z)[s - P_1(z)] .$$

More generally,

$$s^j = Q_{j,1}(z)s^{m-1} + \dots + Q_{j,m}(z) + P(s,z)M_j(s,z) ,$$

where

$$Q_{j+1,k} = \begin{cases} Q_{j,k+1} - Q_{j,1}P_m & \ldots \text{ for } 1 \leq k \leq m-1 \\ -Q_{j,1}P_m & \ldots \text{ for } \quad k = m \end{cases} ;$$

and

$$M_{j+1}(s,z) = sM_j(s,z) + Q_{j,1}(z) .$$

Thus, for $j \geq m$ and $1 \leq k \leq m$, we obtain

$$|Q_{j,k}(z)| \leq C(1+C)^{j-m} ,$$

where $C = C(z)$ is defined in the lemma. (Let us note that $Q_{m,k}(z) = - P_k(z)$.) Furthermore, by using the recurrence formulae given above, we see that

$$|M_j(s,z)| \leq (r+C)^{j-m} \qquad (j \geq m)$$

where $r = \max \{|s|,1\}$ and $M_m = 1$.

Now we write

$$F(s,z) = \sum_{j=0}^{\infty} F_j(z)s^j = \sum_{k=0}^{m-1} (\sum_j F_j(z)Q_{j,k}(z))s^k + (\sum_{j=0}^{\infty} M_j(s,z)F_j(z))P(s,z).$$

The rearrangements of series we just made are admissible because of absolute convergence which follows from the estimates

$$|F_j(z)| \leq 2^{-j}\rho^{-j} \max \{|F(\sigma,z)|: |\sigma| \leq 2\rho\}.$$

Applying these estimates with $\rho = (1+C)$ to the coefficients of s^k, we get

$$|R_j(z)| = |\sum_j F_j(z)Q_{j,k}(z)| \leq 2(1+C) \max \{|F(\sigma,z)|: |\sigma| \leq 2(1+C)\}.$$

Similarly, if $\rho = (r+C)$, we obtain

$$|G(s,z)| = |\sum F_j(z)M_j(s,z)| \leq (r+C)^{-m} \max \{|F(\sigma,z)|: |\sigma| \leq 2(r+C)\},$$

and the lemma is proved.

Remark 4. This proof does not extend immediately to the case when the function F is defined only on an open set $A \times B$, $A \subseteq \mathbb{C}$, unless better estimates are available for $Q_{j,k}$ and M_j. If $P_j(0) = 0$, this can be achieved by sufficiently shrinking the set $A \times B$.

Lemma 2. If $P(s,z)$ is a distinguished polynomial in s of degree m in an open set $\mathbb{C} \times B$, then for every $z \in B$ and $\delta > 0$ there exists δ_1, $\frac{\delta}{2} < \delta_1 < \delta$, and a neighborhood N of z such that

$$|P(s,\zeta)| \geq \frac{1}{2}\delta^m 4^{-m}(m+1)^{-m} \quad \text{for} \quad |s| = \delta_1 \text{ and } \zeta \in N.$$

Proof. Given a point z, we divide the annulus $\delta/2 \leq |s| \leq \delta$ into $m + 1$ equal annuli. Then there must be at least one of them, in which $P(s,z)$ has no roots. Let $D = \{|s| = \delta_1\}$ be the circle passing through the middle of this annulus. If $\alpha_1(z),\ldots,\alpha_m(z)$ are the roots of $P(s,z) = 0$, we have, for $s \in D$,

$$|s-\alpha_i(z)| \geq \frac{\delta}{4(m+1)} .$$

Since $P(s,z) = \prod_{i=1}^{m} (s-\alpha_i(z))$, we get

$$\min_D |P(s,z)| \geq [\delta/4(m+1)]^m .$$

By the continuity of P, there is a neighborhood N of z such that

$$\min_{\zeta\in N;s\in D} |P(s,\zeta)| \geq \frac{1}{2}[\delta/4(m+1)]^m .$$

Corollary. (Ehrenpreis [23]-Malgrange [35]). If $P(z)$ is a polynomial of degree m and $F(z)$ an analytic function in the polydisk $\Delta = \{|z_j| < \delta : j = 1,\ldots,m\}$ such that

$$\max_{z\in\Delta} |P(z)F(z)| \leq M ,$$

then, for some constant C depending only on P,

$$\max_{z \in \Delta_1} |F(z)| \leq C\delta^{-m}$$

with $\Delta_1 = \{|z_j| < \delta/2 : j = 1,2,\ldots,m\}$.

Proof. There is a varaible, say z, for which we can write P as

$$P(z) = a_{m_1}(z_2,\ldots,z_n)z_1^{m_1} + \ldots + a_o(z_2,\ldots,z_n) ,$$

where $a_{m_1}(z_2,\ldots,z_n)$ is a polynomial of degree $m-m_1$. Let z^o be a fixed point in Δ_1. Then for any $z_2,\ldots,z_n$, $|z_j| < \delta$, the previous lemma yields

$$|F(z_1^o,z_2,\ldots,z_n)a_{m_1}(z_2,\ldots,z_n)| \leq M\, 4^{m_1}(m+1)^{m_1}\delta^{-m_1} .$$

The corollary then follows by induction.

Let us return to the local version of the Weierstrass theorem.

<u>Lemma</u> 3. If P(s,z) is a distinguished polynomial in s of degree m such that for some $\delta_1 > 0$ and an open set $N \subseteq \mathbb{C}^n$,

$$\min_{s \in D; z \in N} |P(s,z)| \geq \varepsilon \qquad (D = \{s: |s| = \delta_1\}) ,$$

then for any analytic function F(s,z) on $V = \{s: |s| < \delta_1\} \times N$ we can write

$$F(s,z) = G(s,z)P(s,z) + R(s,z) ,$$

where G(s,z) is analytic on V, and

$$R(s,z) = \sum_{j=0}^{m-1} s^j R_j(z) ,$$

with $R_j(z)$ analytic in N. Moreover, if

$$M = \sup \{|F(s,z)|: (s,z) \in V\} < \infty$$

and

$$K = \sup \{|P_j(z)| : z \in N,\ j = 1,\dots,m\} < \infty ,$$

then

$$|R_j(z)| \le \frac{1}{\varepsilon} MK\delta_1 (1+\delta_1+\dots+\delta_1^{m-1})$$

and

$$|G(s,z)| \le \frac{1}{\varepsilon} M + \frac{1}{\varepsilon} MK\delta_1 (1+\delta_1+\dots+\delta_1^{m-1})^2 .$$

Proof: Let $\delta' < \delta_1$ and

$$G(s,z) = \frac{1}{2\pi i} \int_{|\sigma|=\delta'} \frac{F(\sigma,z)\, d\sigma}{P(\sigma,z)(\sigma-s)} \qquad (|s| < \delta',\ z \in N).$$

Then, if δ' is sufficiently close to δ_1, $G(s,z)$ does not depend on δ'. Hence we may assume that F is analytic for $|s| \le \delta_1$, since the above bounds will then be obtained by taking the limit of bounds valid for $\delta' < \delta_1$. Let

$$R(s,z) \overset{\text{def}}{=} F(s,z) - G(s,z)P(s,z).$$

Let us show that the function R has the desired properties. By the definition of $G(s,z)$,

$$R(s,z) = \frac{1}{2\pi i} \int_{|\sigma|=\delta_1} \frac{F(\sigma,z)}{\sigma-s}\, d\sigma - \frac{1}{2\pi i} P(s,z) \int_{|\sigma|=\delta_1} \frac{F(\sigma,z)\, d\sigma}{P(\sigma,z)(\sigma-s)}$$

$$= \frac{1}{2\pi i} \int_{|\sigma|=\delta_1} \frac{F(\sigma,z)}{P(\sigma,z)} \left[\frac{P(\sigma,z)-P(s,z)}{\sigma-s}\right] d\sigma .$$

However,

$$\frac{P(\sigma,z)-P(s,z)}{\sigma-s} = \sum_{j=0}^{m} P_j(z) \frac{\sigma^j - s^j}{\sigma-s} = \sum_{j=0}^{m} P_j(z)(\sigma^{j-1}+\sigma^{j-2}s+\dots+s^{j-1})$$

$$= \sum_{j=0}^{m-1} \Big(\sum_{i=j+1}^{m} P_i(z)\sigma^{i-j-1}\Big) s^j ,$$

where we have used the equality $P_m(z) \equiv 1$. This shows that $R(s,z)$ is a polynomial in s of degree $\leq m-1$. The estimate for the coefficients $R_j(z)$ also follows. Finally, using the maximum modulus theorem and the analycity of $G(s,z)$, we obtain the estimate for G,

$$\max_{|s|=\delta_1} |G(s,z)| \leq \frac{1}{\varepsilon} \max_{|s|=\delta_1} |G(s,z)P(s,z)|$$

$$\leq \frac{1}{\varepsilon} \max_{|s|=\delta_1} \{|F(s,z)| + |R(s,z)|\}.$$

(Actually, here we should have taken first $\tilde{\delta}_1 < \delta_1$ and then the limit for $\tilde{\delta}_1 \to \delta_1$, etc.)

<u>Remark</u> 5. If all **m** zeroes (counting their multiplicities) of $P(s,z) = 0$, $z \in N$, lie inside the circle $|s| < \delta_1$, then the functions $R_j(z)$ and $G(s,z)$ are <u>uniquely</u> determined. Indeed, if

$$F = GP + R = G'P + R' \qquad (|s| < \delta_1),$$

then $\tilde{R} = R(s,z) - R'(s,z)$ is a polynomial in s of degree at most m-1; but $\tilde{R}$ has at least m zeros (counting multiplicities) in $|s| < \delta_1$ for each $z \in N$, hence $\tilde{R}(w) \equiv 0$. In fact the same reasoning shows that there is a uniqueness in the global decomposition of Lemma 1. Moreover, using Lemma 3 we could have also deduced the global version of Lemma 1 from the local one by taking δ_1 so large that all zeros of $P(s,z) = 0$ would be inside the circle $|s| < \delta_1$ for a fixed z. (Nevertheless the above estimates are slightly better.)

However there is another way of computing the coefficients $R_j(z)$ as shown in the next lemma.

<u>Lemma</u> 4. Let $Q(s)$ be a polynomial in one variable of degree m. Let $s_1,\ldots,s_k$ be all the (distinct) roots of Q with multiplicities $m_1,\ldots,m_k$, $\sum m_j = m$. Then, given arbitrary complex numbers a_{pq}, $p = 1,\ldots,k$, $q = 0,1,\ldots,m_{p-1}$, there exists a unique polynomial

$R(s) = R_1 s^{m-1} + \ldots + R_{m-1} s + R_m$ such that

$$\frac{d^q}{ds^q} R(s_p) = a_{pq} \; . \tag{12}$$

Moreover, $R_j = D_j/D$, for $j = 1,2,\ldots,m$, where the D_j's are linear combinations of the numbers a_{pq} with coefficients which are polynomials in the roots s_p ; and, D is a polynomial in the coefficients of Q. The formulae for D_j and D depend only on the partition $m_1,\ldots,m_k$ of m.

Proof: It suffices to observe that (12) is a system of m linear equations with the coefficients R_j as unknowns. Let D be the square of the determinant of this system. (D is similar to the Vandermonde determinant.) Then

$$D = c \prod_{i \neq j} (s_i - s_j)^{m_i m_j} \tag{13}$$

where c is the positive number [2]

$$c = \left(\prod_{i=1}^{k} \prod_{j=1}^{m-1} \right)^2 .$$

Therefore, $D \neq 0$, and the system (12) has a unique solution which can be found by Cramer's rule. Hence R_j is the quotient $\tilde{D}_j/\sqrt{D}$. The numerator $\tilde{D}_j$ is a linear combination of a_{pq} with coefficients which are $(m-1)\times(m-1)$ determinants involving powers of the roots s_p. Let $D_j \overset{\text{def}}{=} \tilde{D}_j \sqrt{D}$. Then $R_j = D_j/D$. On the other hand, D is a symmetric function of the roots of the polynomial Q(s); and, as such it can be written as a polynomial in the coefficients of Q. The rest of the lemma is clear.

Remarks: 6. If P(s,z) is a distinguished polynomial in s of degree m, then for every fixed z, Lemma 4 gives formulae for D(z) and $D_j(z)$. The expressions for these quantities change from point to point, but it is not difficult to see that there are only a finite number of different systems of such formulae. In particular, $D(z) = p_z(P_1(z),\ldots,P_m(z))$, where p_z is a polynomial. Therefore $D(\zeta) = D_z(\zeta) = p_z(P_1(\zeta),\ldots,P_m(\zeta))$ is an entire function, and we only have a finite number of different D_z's, say $D_{z_1},\ldots,D_{z_t}$. Then,

$$(14) \qquad \Delta(\zeta) = \prod_{j=1}^{t} D_{z_j}(\zeta)$$

is called the <u>discriminant of the function P</u>. Since $\Delta(\zeta)$ is obviously a polynomial in the P_j's, $\Delta(\zeta)$ must be a polynomial in ζ whenever P is a polynomial.

7. Let us compute $\Delta(\zeta)$ for some particular cases. If $P(w) = s^m - P_m(z)$, then $D(z) = c_1(P_m(z))^{m-1}$ when $P_m(z) \neq 0$, and $D(z) = c_2$ otherwise. Hence $\Delta(z) = c(P_m(z))^{m-1}$ for some constant $c \neq 0$. In particular, for $m = 1$, $\Delta(z) = \text{const.} \neq 0$. For $P(w) = s^2 + P_1(z)s + P_2(z)$ we obtain $\Delta(z) = c(P_1^2(z)-4P_2(z))$, etc.

Next we have to know how the functions D_j, D change from one point to another. The answer is given by the following lemma due to A. Ostrowski [40].

<u>Lemma</u> 5. Let $P(s,z)$ be a distinguished polynomial in s of degree m. For each z, let $s_1(z),s_2(z),\ldots,s_m(z)$ denote the roots of $P(s,z) = 0$ counted according to their multiplicities. Then, for any two points z' and z", the roots of $P(s,z') = 0$ and $P(s,z") = 0$ can be rearranged so that for all j,

$$(15) \qquad |s_j(z') - s_j(z")| \leq 6m^2c_1c_2|z'-z"|^{1/m},$$

where

$$
(16)\quad \begin{cases} c_1 = \max\limits_j (1, |P_j(z')|^{1/j}, |P_j(z'')|^{1/j}) , \\ c_2 = \max\limits_j \{|\text{grad } P_j(z)|^{1/m}: z = tz' + (1-t)z'', 0 \le t \le 1\}, \\ |\text{grad } P_j(z)|^2 = \sum\limits_{k=1}^{n} |\frac{\partial}{\partial z_k} P_j(z)|^2 . \end{cases}
$$

Proof: To begin with, let us first observe that all roots of any polynomial

$$
Q(s) = s^m + a_1 s^{m-1} + \ldots + a_m ,
$$

lie in the disk

$$
A = \{s: |s| \le \max_j |ma_j|^{1/j}\}.
$$

Actually, for $s \notin A$,

$$
|Q(s)| \ge |s^m| - |a_1 s^{m-1} + \ldots + a_m| > 0 .
$$

Let us set

$$
(17)\qquad \alpha = \max \{|mP_j(z')|^{1/j}, |mP_j(z'')|^{1/j}: 1 \le j \le m\}
$$

and

$$
(18)\qquad \beta = \left(\sum_{j=1}^{m} |P_j(z') - P_j(z'')|\alpha^{m-j} \right)^{1/m} .
$$

Then clearly $|s_j(z')| \le \alpha$ and $|s_j(z'')| \le \alpha$. Moreover, if

$$
p_t(s) = (1-t)P(s,z') + tP(s,z'') \qquad (0 \le t \le 1)
$$

and $p_t(s_o) = 0$ for some fixed t, then again $|s_o| \le \alpha$.

Let us consider m closed disks of radius β with centers at $s_j(z')$. Let $C_1,\ldots,C_k$ be the different connected components of the union of these disks. In this way all roots $s_j(z')$ can be divided into k groups of roots lying in the same set C_ℓ. Let $\beta' > \beta$ be chosen so close to β that

(i) $(2m-1)\beta' < 2m\beta$; and,

(ii) the disks of radius β' with centers at $s_j(z')$ define the same number k of connected components $C_1', \ldots, C_k'$.

We may also assume that $C_i \subseteq C_i'$. Let Γ_i be the boundary of C_i'. Then Γ_i consists of a finite number of circular arcs. We claim that for all t and i,

(19) $$p_t(s) \neq 0 \quad \text{on} \quad \Gamma_i \; .$$

Let us assume the contrary, i.e. suppose that for some t and i, $p_t(s_o) = 0$ for some $s_o \in \Gamma_i$. By the construction of the curves Γ_i, for all j we have,

(20) $$|s_o - s_j(z')| \geq \beta' \; .$$

On the other hand, by (17) and (19),

$$\prod_{j=1}^{m} |s_o - s_j(z')| = |P(s_o,z')| = |P(s_o,z') - p_t(s_o)|$$

$$= t|P(s_o,z') - P(s_o,z'')| \leq t\beta^m ;$$

thus, for at least one j we must have $|s_o - s_j(z')| < t^{1/m}\beta \leq \beta < \beta'$, which contradicts condition (20). Hence, by Rouché's theorem, the polynomials $P(s,z') = p_o(s)$ and $P(s,z'') = p_1(s)$ have the same number of zeros, say ρ_i, inside the region C_i'. Then, however,

(21) $$|s_i(z') - s_j(z'')| \leq (2\rho_i - 1)\beta' < 2m\beta$$

for all $s_i(z')$, $s_j(z'') \in C_i'$ (cf. the construction of the regions C_i'). Therefore we reorder the roots $s_j(z')$ and $s_j(z'')$ in such a way that the first ρ_1 roots will lie in C_1, etc.; and, furthermore, so that (21) will hold with i = j for all indices. The estimate (15) then follows.

<u>Corollary 1</u>. For $z \in \mathbb{C}^n$ and $r > 0$, let

(22) $$M(\zeta,r) = 6m^2 \max\{1, |P_j(\zeta)|^{1/j}: |z-\zeta| \le r,\ j = 1,\dots,m\}$$

$$\times \max\{|\mathrm{grad}\ P_j(\zeta)|^{1/m}: |z-\zeta| \le r,\ j = 1,\dots,m\}$$

(cf. (16)), and

(23) $$\rho(z,r) = \max\{1+|mP_j(\zeta)|^{1/j}: |z-\zeta| \le r,\ j = 1,\dots,m\}.$$

Then by Lemma 5, for all $z,z'' \in \{\zeta: |z-\zeta| \le r\}$, all the roots $s_i(z')$ ($s_j(z'')$) of $P(s,z') = 0$ ($P(s,z'') = 0$ resp.) lie inside the disk $|s| \le \rho(z,r)$; and, the roots can be reordered so that, for all j,

(24) $$|s_j(z')-s_j(z'')| \le M(z,r)|z'-z''|^{1/m} .$$

§2. PROOF OF THE THEOREM

Definition 1. An analytic variety V in $\mathbb{C}^p$ is the set of common zeros of a finite collection of entire functions $F_1,\ldots,F_r$. A set V is called a Zariski variety ($\mathcal{Z}$-variety), if V is of the form $V = V'-V''$ where V', V'' are analytic varieties. A multiplicity variety $\mathcal{V}$ is a finite collection $\{V_1,d_1;\ ..;V_r,d_r\}$, where the V_j's are $\mathcal{Z}$-varieties (not necessarily distinct), and the d_j's are certain differential operators in $\mathbb{C}^p$ with constant coefficients, i.e. each d_j is a linear combination of $\partial^\alpha/\partial z^\alpha$, $\alpha = (\alpha_1,\ldots,\alpha_r)$. If G is an open (or closed) subset of $\mathbb{C}^p$, then $\mathcal{V} \cap G$ denotes the collection $\{V_1 \cap G, d_1;\ \ldots;\ V_1 \cap G, d_r\}$. If H is an entire function, then $H|\mathcal{V}$, i.e. the restriction of H to $\mathcal{V}$, is defined as the collection of functions $\{H_j\}$, where each H_j is defined on V_j by

$$H_j = d_jH|V_j = \text{the restriction of } d_jH \text{ to } V_j\ .$$

Similarly, for H analytic in G, we can define the restriction of H to $\mathcal{V} \cap G$. Conversely, a collection of functions H_j, H_j defined on V_j, is called an analytic function on $\mathcal{V}$, if there exists an entire function H satisfying $H|\mathcal{V} = \{H_j\}$. Similarly one can define analytic functions on $\mathcal{V} \cap G$, etc.

Let W be an AU-space with base U and an AU-structure $\mathcal{K} = \{k\}$. Furthermore, let $\mathcal{V}$ be a fixed multiplicity variety. If $H \in \hat{U}$, $k \in \mathcal{K}$ are arbitrary, then for some constant $c > 0$, $|H(z)| \leq c\ k(z)$; and, obviously, for all j and z,

$$(25) \qquad |d_jH(z)| \leq c' \max\ \{k(z'):\ |z'-z| \leq 1\} \quad (c'=\text{const.}).$$

By Def. 1, I (cf. (iv)), the right-hand side of (25) is also a majorant in $\mathcal{K}$. Hence we can define the space $\hat{U}(\mathcal{V})$ as the set of all analytic functions $\{H_j\}$ on $\mathcal{V}$, satisfying for any $k \in \mathcal{K}$,

$$(26) \qquad |H_j(z)| = \mathcal{O}(k(z)) \qquad (z \in V_j;\ \forall j)$$

It is clear that condition (26) defines not only the space $\hat{U}(\mathfrak{N})$ as a set, but also it defines an l.c. topology on $\hat{U}(\mathfrak{N})$, under which the natural mapping

$$(27) \qquad \iota : \hat{U} \to \hat{U}(\mathfrak{N}) \quad \text{where} \quad \iota(H) = H|\mathfrak{N}$$

is continuous. The following notation will be used

$$d_j'H(z) = \begin{cases} d_jH(z) & \ldots \text{ for } z \in V_j \\ 0 & \ldots \text{ for } z \notin V_j \end{cases}$$

$$\|H(z)\| = \|H(z)\|_{\mathfrak{N}} = \sum_j |d_j'H(z)|.$$

Furthermore, we shall say that an entire function H is in $\hat{U}(\mathfrak{N})$, if the analytic function $\{H_j\}$, defined on $\mathfrak{N}$ by H, is in $\hat{U}(\mathfrak{N})$, etc.

<u>Theorem 1</u> (The fundamental principle). Let W be a PLAU-space of dimension n+1 with base U and an AU-structure $\mathcal{K}$. Let P(s,z) be a distinguished polynomial in s of degree m. Assume that

(i) $\mathcal{C}_P$ is a convolutor of W *

(ii) For every $k \in \mathcal{K}$ there exists $k' \in \mathcal{K}$ such that

$$(28) \qquad M(z,1)\rho(z,1)k'(s,z) \leq k(s,z) \quad \text{for all } s,z \;{}^{**}$$

Then there exists a multiplicity variety $\mathfrak{N}$ such that

(I) $H \in \hat{U}$, $H|\mathfrak{N} = 0$ if and only if $H = PG$ for some $G \in \hat{U}$;

and,

(II) for any $H \in \hat{U}(\mathfrak{N})$ there exists a function $F \in \hat{U}$ such that

* i.e. the map $H \mapsto PH$ of $\hat{U} \to \hat{U}$ is continuous; cf. Def. 2, I.
**cf. (22), (23).

(a) $\Delta(z)H(s,z) = F(s,z) + P(s,z)G(s,z)$

for some entire function G *

(b) $\Delta H|\mathcal{V} = F|\mathcal{V}$; and,

(c) the mapping $\kappa : \hat{U}(\mathcal{V}) \to \hat{U}$ given by $H \mapsto F$ is continuous.

Moreover, suppose that, for each $z_o \in \mathbb{C}^n$ and b positive, there exists a constant $Q(z_o,b)$ such that, for each $\gamma > 0$ and F analytic in $S(w_o) = \{w: |s-s_o| \leq \gamma, |z-z_o| \leq b\}$,

$$\max \{|F(w)|: w \in \tfrac{1}{2} S(w_o)\} \leq Q(z_o,b) \max \{\Delta(z)F(z): S(w_o)\}.$$

Furthermore, for $d > 0$, let

$$\tilde{Q}(z_o,d) \overset{\text{def}}{=} \max \{Q(z,b(z)): |z-z_o| \leq d\},$$

and

$$b(z) = \min \left\{\left(\frac{d}{M(z,d)}\right)^m, d\right\}$$

Finally we shall assume that

(ii') for every $k \in \mathcal{K}$ there exists a $k' \in \mathcal{K}$ such that $\tilde{Q}(z,d)M(z,2d)\rho(z,2d)k'(w) \leq k(w)$.

Then,

(III) the l.c. spaces $\hat{U}/P\hat{U}$ and $\hat{U}(\mathcal{V})$ are isomorphic.

Remark 8. The most natural method of proving this theorem seems to be be the following. First, the multiplicity variety $\mathcal{V}$ should be defined in terms of pieces of the analytic variety $V_P = \{w: P(w) = 0\}$. Then given any $H \in \hat{U}(\mathcal{V})$ we could apply Lemma 1 and Lemma 4 (the latter one is usually called the Lagrange interpolation formula), and obtain the function $R(s,z)$ with the desired properties (in particular, $R(w)$ will be entire and a distinguished polynomial in s).

* Δ is the discriminant of P, cf. (14).

Moreover, we know how to compute the coefficients $R_j(s,z)$ in R in terms of the "values" of H on $\mathcal{W}$. However, there is one difficulty in this approach; namely, the coefficients of R(s,z), for a given $(s,z) \in V$, are computed in terms of the values of H at certain points $(s',z) \in V$ with values s' far away from the original value s. Therefore we are not able to deduce that $R(s,z) \in \hat{U}$. The way out of this difficulty consists in introducing a convenient covering of $\mathbb{C}^{n+1}$ and imposing certain (local) restrictions on the functions defined on the elements of the covering. Next one has to show that a certain cohomology group vanishes. The procedure of extending a function $H \in \hat{U}(\mathcal{W})$ to $\mathbb{C}^n$ can be decomposed into several steps. First, we extend H from $\mathcal{W}$ to special rectangles. Then these extensions are pasted together using the Cartan-Oka-Serre procedure [25] to obtain a cocycle satisfying good bounds. Finally, one has to repeat this procedure this time with due care to bounds (at this stage the PLAU-structure enters into the proof). The result will be a function in $\hat{U}$ having all the properties prescribed by the theorem.

Proof: There are several possible choices for the definition of the multiplicity variety $\mathcal{W}$. For our purposes the simplest and the most natural choice will be sufficient. Let us call $\mathcal{W}$ the multiplicity variety

$$(29)\quad \begin{cases} V_1 = \{w : P(w) = 0\} \setminus \{w : \frac{\partial}{\partial s} P(w) = 0\}; & \text{operators d: identity;} \\ V_2 = \{w : P(w) = \frac{\partial P}{\partial s}(w)\} \setminus \{w : \frac{\partial^2}{\partial s^2} P(w) = 0\}; & \text{operators d: identity, } \frac{\partial}{\partial s}\text{;} \\ \vdots & \\ V_m = \{w : P(w) = \frac{\partial P}{\partial s}(w) = \ldots = \frac{\partial^{m-1} P(w)}{\partial s^{m-1}} = 0\}; & \text{operators: identity, } \\ & \frac{\partial}{\partial s}, \ldots, \frac{\partial^{m-1}}{\partial s^{m-1}}. \end{cases}$$

Obviously, $\bigcup_j V_j = V_P = \{w: P(w) = 0\}$ and $PG|\mathcal{V} = 0$ for any entire function G. For every z fixed, the decomposition of Lemma 1 is completely determined by the "values of the function H on the variety $\mathcal{V}$ above the point z". In particular, if H is entire and $H|\mathcal{V} = 0$, then $H(w) = P(w)Q(w)$ for some entire Q; and, by Lemma 2, $H \in \hat{U} \Rightarrow Q \in \hat{U}$. Thus we already know that the mapping

$$\mu: \hat{U}/P\hat{U} \to \hat{U}(\mathcal{V}) \tag{30}$$

is injective and continuous. Consider the mapping $\pi: \hat{U} \to \hat{U}/P\hat{U}$ and $\tilde{\Delta}: H \to \Delta H$. Then we are supposed to prove that in the diagram

$$\begin{array}{ccc} \hat{U} & \xrightarrow{\ \pi\ } & \hat{U}/P\hat{U} \\ \kappa \nwarrow \searrow \iota & & \swarrow \mu \\ & \hat{U}(\mathcal{V}) & \end{array} \tag{31}$$

the mapping κ is continuous, μ injective and $\mu \circ \pi \circ \kappa = \tilde{\Delta}$. The theorem does not state that μ is surjective (cf. part (II) of the theorem) unless we suppose more about the discriminant Δ in part (II) which guarantees that $\tilde{\Delta}$ is invertible.

Let us fix a point $w_0 = (s_0, z_0) \in \mathbb{C}^{n+1}$ and a constant $a > 0$. Corollary 1 of Lemma 5 says that if

$$b \leq \min\{a, [a/M(z_o;a)]^m\}, \tag{32}$$

then, for each z such that $|z-z_o| \leq b$, the roots $s_k(z)$ of $P(s,z) = 0$ lie in the disks $|s-s_k(z_o)| \leq a$. Let $T(z_o)$ be the union of these circles and $T_k(z_o)$ the connected component containing $s_k(z_o)$. Then either $T_k(z_o) = T_j(z_o)$ or $T_k(z_o) \cap T_j(z_o) = \emptyset$. Given any pair of numbers c,d such that $0 < c < d$, we can find $\gamma = \gamma(s_o, z_o, c, d)$ such that

$$d < \gamma < 2(d + mc + ma) . \tag{33}$$

Moreover, if the disk $|s-s_o| < \gamma$ intersects some $T_k(z_o)$, then

$$T_k(z_o) \subseteq \{s: |s-s_o| < \gamma - c\}.$$

If F is any function analytic in $\{w: |s-s_o| < \gamma, |z-z_o| < b\}$, we claim that there exists an analytic function F' on $\mathcal{V} \cap \{w: |z-z_o| < b\}$ such that

$$(34) \qquad \begin{cases} F'|\mathcal{V} = F|\mathcal{V} & \text{for } w \in \{w: |s-s_o| < \gamma, |z-z_o| < b\} \\ F'|\mathcal{V} = 0 & \text{for } w \in \{w: |s-s_o| \geq \gamma, |z-z_o| < b\} \end{cases}$$

To verify (34), let $\chi(s) \in C_o^\infty(\mathbb{R})$ where $\chi \equiv 1$ for $|s-s_o| \leq \gamma - c$ and $\chi \equiv 0$ for $|s-s_o| \geq \gamma$. Then χF can be extended as 0 to the rest of the strip $Z(z_o,b) = \{w: |z-z_o| < b\}$ and becomes a C^∞-function in $Z(z_o;b)$. Then F' is defined by $F' = \chi F + uP$, where u is the solution of

$$(35) \qquad \bar{\partial}u = -\frac{\bar{\partial}(\chi F)}{P} .$$

Obviously the right-hand side of (35) is a C^∞-function in $Z(z_o,b)$ and in the set $|s-s_o| \leq \gamma - c$ and $|s-s_o| \geq \gamma$. By Th. 4.4.3 of [29] we know that a solution u to (35) exists; in particular, u will be analytic in the strip $Z(z_o,b)$ except for the set $\{w: \gamma-c \leq |s-s_o| \leq \gamma, |z-z_o| \leq b\}$ Now it follows that conditions (34) are satisfied. To summarize, we have found the following: If $S(w_o) \overset{\text{def}}{=} \{w: |s-s_o| < \gamma, |z-z_o| < b\}$, then there exists a function F' analytic in $Z(z_o,b)$ such that

$$(36) \qquad F|\mathcal{V} \cap S(w_o) = F'|\mathcal{V} \cap S(w_o)$$

and

$$(37) \qquad F(w) = F'(w) + P(w)G(w)$$

where G is analytic in $S(w_o)$.

Now we can apply the Lagrange interpolation formula (i.e. Lemmas 1,4) to the function F' and obtain the function R(w) which is a distinguished polynomial in s of degree m-1, defined in $Z(z_o,b)$, such that the coefficients $R_j(z)$ depend linearly on the values $F'|\mathcal{V}$.

Since $F'|\mathfrak{N} = 0$ outside the rectangle $S(w_o)$, the R_j's depend only on $F'|\mathfrak{N} \cap S(w_o)$. In other words, given $H \in \hat{U}(\mathfrak{N})$ and constants a,c,d, for every $w_o = (s_o, z_o)$, there exist numbers b and γ, satisfying (32) and (33) such that in $S(w_o)$ we have

(38) $$H(w) = R_{w_o}(w) + P(w)Q_{w_o}(w)$$

and

(39) $$\max_{w \in S(w_o)} |\Delta(z)R_{w_o}(w)| \leq C\rho(z_o,b)^{\ell} \max_{w \in S(w_o)} \|H(w)\|$$

where ℓ and C are positive numbers independent of w_o and b. Furthermore, we claim that there are functions F_{w_o}, G_{w_o} analytic in $\{w: |s-s_o| < \frac{d}{2}, |w-w_o| < \frac{d}{2}\}$ such that

(40) $$\Delta(z)H(w) = F_{w_o}(w) + P(w)G_{w_o}(w) ,$$

and

(41) $$|F_{w_o}(w)| \leq CM(z_o,2d)\rho(z_o,2d)^{\ell} \max \{\|H(w')\|: w' \in S'(d;w_o)\},$$

where $S'(d,w_o) = \{w: |s-s_o| < 2(m+1)d+m, |z-z_o| < d\}$. (Here we used (33) with $d = a = c = 1/2$). The proof of the existence of F_{w_o} and G_{w_o} follows along the same lines - namely the Cartan-Oka-Serre method - as the proof of (43,p) => (44,p) given below. Actually, here this procedure becomes even simpler than in the latter case, because it is finite; and, thus the convergence factors ϕ,ψ which appear in the proof of (43,p) => (44,p) are not necessary. Therefore, the proof of the existence of the functions F_{w_o} and G_{w_o} can be omitted.

Let us write $w = (x_1, x_2, \ldots, x_{2n-2})$, $z_j = x_{2j-1} + ix_{2j}$ for $j = 1,2,\ldots,n$ and $s = x_{2n+1} + ix_{2n+2}$. For $p = 1,2,\ldots,2n+3$, $\alpha \in \mathbb{R}^{2n+2}$ and $\delta > 0$, we set

$$B_p(\alpha,\delta) = \{w \in \mathbb{C}^{n+1}: |x_j - \alpha_j| < \delta \text{ for } j \geq p\}.$$

Since the first (p-1) coordinates of α do not appear in the definition of α, they can be taken to be zero. $B_1(\alpha,\delta)$ is just the "cube" of center α and side 2δ, and $B_{2n+3} = \mathbb{C}^{n+1}$ for all α,δ.

Let us choose $\underline{d}$ rather large, say $d = 100n$, and let $L = \{\alpha\}$ be the set of all lattice points in $\mathbb{R}^{2n+2}$. Then there is a system $\{F_\alpha\}$ of analytic functions associated with the system of rectangles $\{B(\alpha, \frac{d}{2})\}_{\alpha\varepsilon L}$ as described above (cf. (40),(41)). The system $\{F_\alpha\}$ is a cocycle in the sense that, for any pair $\alpha,\alpha' \varepsilon L$, there is a function $G_{\alpha,\alpha'}$ analytic in $B_1(\alpha,\frac{d}{2}) \cap B_1(\alpha',\frac{d}{2})$ such that $F_\alpha - F_{\alpha'} = PG_{\alpha,\alpha'}$. Moreover the functions F_α satisfy good estimates.
The idea of the following proof is to extend the functions F_α to sets B_p as p increases (and, of course, going from p to p+1, i.e. extending by one real variable at a time). Finally we will end up with one function F satisfying good estimates in $B_{2n+3} = \mathbb{C}^{n+1}$, i.e. $F \varepsilon \hat{U}$.

Two lattice points α,α' will be called p-<u>semiadjacents</u>, if

(42) $\alpha_j = \alpha_j' = 0$ for $j=1,\ldots,p-1$; and, $\alpha_i = \alpha_j'$ for $j=p+1,\ldots,2n+2$.

If $\alpha_p < \alpha_p'$, we shall write $\alpha < \alpha'$. The points α,α' will be called p-<u>adjacent</u>, if, in addition to (42), $|\alpha_p - \alpha_p'| = 1$. Moreover, the following spaces will be needed: For $\delta > 1$, we define

$$\mathcal{O}_p(\delta) = \{F = \{F_\alpha\} : F_\alpha \text{ analytic in } B_p(\alpha,\delta)\};$$

$$\hat{U}_p(\delta) = \{F = \{F_\alpha\} \varepsilon \mathcal{O}_p(\delta) : \lim_{|w|\to\infty} \frac{F_\alpha(w)}{k(w)} = 0 \text{ for all } k \varepsilon \mathcal{K}\};$$

$$\hat{U}_p(\delta,P) = \{F=\{F_\alpha\}\varepsilon\hat{U}_p(\delta) : F_\alpha - F_{\alpha'} = PG_{\alpha,\alpha'} \text{ on } B_p(\alpha,\delta)\cap B_p(\alpha',\delta) \text{ with } G_{\alpha,\alpha'} \text{ analytic for all } \alpha,\alpha' \varepsilon L\};$$

$$\hat{U}^*_p(\delta,P) = \{F=\{F_\alpha\}\varepsilon\hat{U}_p(\delta) : F_\alpha - F_{\alpha'} = PG_{\alpha,\alpha'} \text{ on } B_p(\alpha,\delta)\cap B_p(\alpha',\delta) \text{ with } G_{\alpha,\alpha'} \text{ analytic for all p-semiadjacent } \alpha,\alpha' \text{ in } L\}.$$

All four spaces are equipped with the corresponding natural topologies. For a small positive ε, one can define the natural maps

$$\lambda_p : \hat{U}_{p+1}(\delta, P)/(P\hat{U}_{p+1}(\delta-\varepsilon) \cap \hat{U}_{p+1}(\delta)) \to \hat{U}_p(\delta, P)/(P\hat{U}_p(\delta-\varepsilon) \cap \hat{U}_p(\delta))$$

which are clearly continuous. Our aim is to prove:

(43,p) $\begin{cases} \text{For } \varepsilon > 0 \text{ small and } F \in \hat{U}_p(\delta) \cap P\,\mathcal{O}_p(\delta), \text{ there exists a} \\ G \in \hat{U}_p(\delta-\varepsilon) \text{ such that } F = PG, \text{ and the mapping } F \mapsto G \text{ is} \\ \text{continuous.} \end{cases}$

(44,p) $\begin{cases} \text{For } \delta \text{ large there exist an } \varepsilon > 0 \text{ and a mapping} \\ \mu_p : \hat{U}_p^*(\delta, P)/(P\hat{U}_p(\delta-\varepsilon) \cap \hat{U}_p(\delta)) \\ \to \hat{U}_{p+1}(\delta-\varepsilon, P)/(P\hat{U}_{p+1}(\delta-2\varepsilon) \cap \hat{U}_{p+1}(\delta-\varepsilon)) \\ \text{such that } \mu_p \text{ is continuous and } \lambda_p \circ \mu_p = \text{identity.} \end{cases}$

(45,p) λ_p is injective .

The proof by induction will be shown as follows:

(a) (43,p) => (44,p)

(b) (45,p) holds for all p

(c) (43,p) => (43,p+1)

We know that (43,1) holds (cf. Lemma 2). Then (a),(b),(c) will imply that the mapping

$$\nu : \hat{U}_{2n+3}(\delta, P)/(P\hat{U}_{2m+3}(\delta-\varepsilon) \cap \hat{U}_{2n+3}(\delta)) = \hat{U}/P\hat{U}$$

$$\to \hat{U}_1(\delta, P)/(P\hat{U}_1(\delta-\varepsilon) \cap \hat{U}_1(\delta)) ,$$

obtained by composing the mappings λ_p , μ_p , is an isomorphism (in this procedure the numbers δ and ε are modified a finite number of times). In particular we shall obtain functions $F \in \hat{U}$ and G such that, for all j,

(46) $$d_j'(F(w)) = d_j'(\Delta(z)H(w)) ,$$

i.e.,

(47) $$\Delta(z)H(w) = F(w) + P(w)G(w)$$

and this will complete the proof of assertion (III) of the theorem.

Proof of (a): Let $H \in \hat{U}_p^*(\delta, P)$. For α, β, which are p adjacent, we set

$$F_\alpha = H_\beta - H_\alpha .$$

Then $\{F_\alpha\} \in \hat{U}_p(\delta-1) \cap P\,\mathcal{O}_p(\delta-1)$; and, by (43,p), $F_\alpha = PN_\alpha$, $\{N_\alpha\} \in U_p(\delta-1-\varepsilon)$, and $\{F_\alpha\} \mapsto \{N_\alpha\}$ is continuous.

First let us assume that p is odd; then for $p' = (p+1)/2$ we consider in the plane of this p'-coordinate a square with center $\alpha_p+i\alpha_{p+1}$, side $2(\delta-1-\varepsilon)$ and boundary $\Gamma(\alpha)$. Let $\Gamma^+(\alpha)$ be the part of $\Gamma(\alpha)$ in the halfspace $\{w: \mathrm{Re}\, w \leq \alpha_p\}$ and $\Gamma^-(\alpha)$ the other part. If $\phi(\alpha, z_{p'})$ (or $\phi(\alpha, s)$ when $p' = n+1$) is an entire function without zeros inside this square or on $\Gamma(\alpha)$, then the Cauchy formula yields

$$N_\alpha(w) = \frac{\phi(\alpha,z_{p'})}{2\pi i} \int_{\Gamma(\alpha)} \frac{N_\alpha(z_1,\ldots,t,z_{p'+1},\ldots,s)}{(t-z_{p'})\phi(\alpha,t)}\, dt$$

$$= \frac{\phi}{2\pi i}\int_{\Gamma^+} + \frac{\phi}{2\pi i}\int_{\Gamma^-} = N_\alpha^+ - N_\alpha^- .$$

If z_p' is in the closed square with center $\alpha_p+i\alpha_{p+1}$ and side $2(\delta-2-\varepsilon)$, and $t \in \Gamma(\alpha)$, then $|t-z_{p'}| \geq 1$. Then for fixed α, and α' p-semi-adjacent to α, $\alpha' \geq \alpha$, we get

$$|N_{\alpha'}^-(w)| \leq 8\delta\, |\phi(\alpha',z_p)|\, \frac{\max\limits_{t\in\Gamma(\alpha')} |N_{\alpha'}(\ldots)|}{\min\limits_{t\in\Gamma(\alpha')} |\phi(\alpha',t)|} .$$

For any $z_{p'} \in \mathbb{C}$, there are finitely many lattice points α' (bounded independently of z_p') such that $z_{p'}$ is in the interior of $\Gamma(\alpha')$. Let us define

$$m(w) = \max_{\alpha'} \max_{t\in\Gamma(\alpha')} |(1+|t|^2)N_{\alpha'}(z_1,\ldots,z_{p'-1},t,z_{p'+1},\ldots,z_n,s)|,$$

where the max is taken only over those points α' for which the corresponding $z_{p'}^{\alpha'}$ is in the interior of $\Gamma(\alpha')$. Since $\{N_\alpha\}$ is in $\hat{U}_p(\delta-1-s)$, $m(w)/k(w) \to 0$ for all $k \in \mathcal{K}$. Hence we can find an $\tilde{m}$ in the BAU-structure $\mathcal{M}$ of W such that $m \le \tilde{m}$. By Def. 3, I we can assume in the sequel that $n = 1$. By condition (viii) of the same definition, for each $\tilde{m} \in \mathcal{M}$, there exists $m^* \in \mathcal{M}$ such that for every $\alpha \in L$, there is an entire function $\phi(\alpha,z_{p'})$ for which

$$\frac{\tilde{m}(\alpha_p+i\alpha_{p+1})\,|\phi(\alpha,z_{p'})|}{\min_{t\in\Gamma(\alpha)} |\phi(\alpha,t)|} \le m^*(z_{p'}) .$$

In particular,

$$|N^-_{\alpha'}(w)| \le 8\delta m^*(w)(1+|\alpha'_p|^2)^{-1}$$

for $\alpha' \ge \alpha$ and $z_{p'}$ in the square with $\alpha+i\alpha_{p+1}$ and side $2(\delta-2-\varepsilon)$. Similar estimates hold for $N^+_{\alpha''}(w), \alpha'' < \alpha$. Using the indicated estimates we obtain the uniform convergence of the following series,

$$\sum_{\alpha'\ge\alpha} |P(w)||N^-_{\alpha'}(w)| \le 8\delta|P(w)|m^*(w) \sum_{j=0}^{\infty} \frac{1}{1+j^2}$$

$$\sum_{\alpha''<\alpha} |P(w)||N^-_{\alpha''}(w)| \le 8\delta|P(w)|m^*(w) \sum_{j=0}^{\infty} \frac{1}{1+j^2}$$

for $|x_p-\alpha_p| \le \delta-2-\varepsilon$, $|x_{p+1}-\alpha_{p+1}| \le \delta-2-\varepsilon$. (The convergence estimates are independent of α.) Hence we can define an analytic function $\tilde{H}_\alpha$ in $B_p(\alpha,\delta-2-\varepsilon)$ by the formula

$$\tilde{H}_\alpha(w) = H_\alpha(w) + \sum_{\alpha'\ge\alpha} PN^-_{\alpha'}(w) - \sum_{\alpha''<\alpha} PN^-_{\alpha''}(w) .$$

It follows that $\{\tilde{H}_\alpha\} \in \hat{U}_p(\delta-2-\varepsilon)$ and $\tilde{H}_\alpha - H_\alpha \in P\hat{U}_p(\delta-2-\varepsilon)$. Moreover, if β is p-adjacent to α, say $\alpha < \beta$, then in the intersection of their respective domains we obtain from the definition of N_α

$$\tilde{H}_\alpha - \tilde{H}_\beta = H_\alpha - H_\beta + \sum_{\alpha''\geq\alpha} PN^-_{\alpha'} - \sum_{\beta'\geq\beta} PN^-_{\beta'} - \sum_{\alpha''<\alpha} PN^+_{\alpha''} + \sum_{\beta''<\beta} PN^+_{\beta''}$$

$$= H_\alpha - H_\beta + PN^-_\alpha + PN^+_\alpha = H_\alpha - H_\beta + PN_\alpha = 0 .$$

Therefore $\tilde{H}_\alpha$ does not depend on the p-th coordinate of α and $\{\tilde{H}_\alpha\} \in U_{p+1}(\delta-2-\varepsilon)$.

For p even, the proof is the same except that in the definition of N^+_α, N^-_α one has to use strips parallel to the real axis and then apply the second part of condition (viii) of Def. 3, I.

Proof of (b): Suppose that $H \in \hat{U}_{p+1}(\delta,P)$ and $\lambda_p H \in P\hat{U}_p(\delta-\varepsilon)$, then there is $N \in \hat{U}_p(\delta-\varepsilon)$ such that $\lambda_p H = PN$. If β is p-adjacent to α, then in $B_p(\alpha,\delta-\varepsilon) \cap B_p(\beta,\delta-\varepsilon)$,

$$P(N_\beta - N_\alpha) = \lambda_p H_\beta - \lambda_p H_\alpha = \lambda_p(H_\beta - H_\alpha) = 0 .$$

But $H \in \hat{U}_{p+1}(\delta,P)$ implies $N_\beta - N_\alpha = 0$. Therefore $N \in \hat{U}_{p+1}(\delta-\varepsilon)$ which shows that λ_p is injective.

Proof of (c): The same proof as in (a).

To conclude the proof of the theorem, we have to prove part (III) of the theorem, i.e., to show how to remove the discriminant $\Delta(z)$ from (47). It suffices to see that under our hypotheses, inequality (39) implies

$$(48) \qquad \max_{2w\in S(w_o)} |R_{w_o}(w)| \leq Q(z_o,b) \max_{w\in S(w_o)} [\rho(z)^\ell \|H(w)\|] ,$$

where $S(w_o)$ and $Q(z_o,b)$ are defined in part (III) of the theorem. Then (40) and (41) become

$$(40') \qquad H(w) = F_{w_o}(w) + P(w)G_{w_o}(w) ,$$

$$(41') \quad |F_{w_o}(w)| \leq CM(z_o;2d)\rho(z_o,2d)^\ell \tilde{Q}(z_o,d)\max\{\|H(w')\| : w' \in s'(d,w_o)\}$$

and the rest of the proof is the same. Hence the theorem is completely proved.

Corollary A. Assume that W, P and Δ satisfy conditions (i) and (ii) of the theorem. Let $\mathcal{C}_P$ and $\mathcal{C}_\Delta$ be the convolutors corresponding to the multipliers P and Δ, respectively (cf. Def. 2, I). Then every solution f in W of the equation $\mathcal{C}_P(f) = 0$ can be written as

$$(49) \qquad \mathcal{C}_\Delta(f)(y) = \int_{\mathfrak{M}} e^{i\langle y,w\rangle} \frac{d\vec{\nu}(w)}{k(w)} \qquad (y \in \mathbb{R}^{n+1}) ,$$

where $d\vec{\nu} = (d\nu_1,\ldots,d\nu_n)$ are Radon measures with supp $\nu_j \subseteq V_j$ (cf.(29)), and k is a majorant in $\mathcal{K}$. In other words, for each $H \in \hat{U}$ (or $H \in \hat{U}(\mathfrak{M})$),

$$(50) \qquad \langle H, \mathcal{C}_\Delta(f)\rangle = \sum_{j=1}^{r} \int_{V_j} d_j H(w) \frac{d\nu_j(w)}{k(w)} .$$

Proof: As we have seen above, for $H \in \hat{U}$, we have

$$DH = R + PG \ ; \quad G \in \hat{U}$$

and $H \to 0$ in $\hat{U}(\mathfrak{M})$ implies $R \to 0$ in $\hat{U}$. Then, however,

$$\langle H, \mathcal{C}_\Delta(f)\rangle = \langle \Delta H, f\rangle = \langle f, R\rangle \to 0$$

and the Hahn-Banach theorem yields the desired representation (cf. §1, I).

Similarly we obtain,

Corollary B. Under the same hypotheses as in the foregoing corollary, but with (ii) replaced by (ii'), we can write every solution $f \in W$ of $\mathcal{C}_p(f) = 0$ as

$$(51) \qquad f(y) = \int_{\mathfrak{M}} e^{i\langle y,w\rangle} \frac{d\vec{\nu}(w)}{k(w)} .$$

In conclusion let us mention two examples of functions P to which the fundamental principle applies:

(A) P polynomial Then Δ is a polynomial of degree h. By Lemma 2,

$$Q(z,b) \leq C/b^{h+1} .$$

Therefore the quantities $M(z,d)$, $Q(z,d)$ and $\rho(z,d)$ are bounded by $C_d(1+|z|)^N$ for some positive C_d and N, and hypothesis (ii') follows. In particular, this solves the problem mentioned in the beginning of this chapter.

(B) The P_j's are exponential polynomials Let us recall that $F(z)$, $z \in \mathbb{C}^n$, is an exponential polynomial, if F can be written as

$$F(z) = \sum_{k=1}^{m} a_k(z) e^{<\alpha_k, z>}$$

where $a_k(z)$ are polynomials and α_k are complex numbers called the frequencies of F. If the P_j's in Def. 1 are exponential polynomials, then Δ is also an exponential polynomial. Moreover, if all the frequencies of the P_j's are real (or pure imaginary), the same holds for Δ.

The following estimate, generalizing the corollary to Lemma 2, can be derived for any exponential polynomial F. Let us set

(52) $$h_F(z) = \max_k \operatorname{Re} <\alpha_k, z> .$$

Then there exists a polynomial A(t) with positive coefficients depending only on the exponential polynomial F such that, for arbitrary $z_o \in \mathbb{C}^n$, $\varepsilon > 0$ and g entire,

(53) $$e^{h_F(z_o)} |g(z_o)| \leq A(\tfrac{1}{\varepsilon}) \max_{|z-z_o| \leq \varepsilon} |F(z)g(z)| .$$

From here and the definitions of the expressions M, Q and ρ, it follows that all these expressions can be estimated by

$$\text{const.}\ (1+|z|)^N e^{h(z)} \quad ,$$

where $h(z) = \max_k \mathrm{Re} < \beta_k, z >$, and β_k's are complex vectors depending on P. Let us observe that if all the frequencies α_k are real (or pure imaginary), then the same holds for the vectors β_k. For instance, in the case $W = \mathcal{D}'_\omega$ we have to take all α_k's pure imaginary (otherwise P would not be a multiplier in the corresponding $\hat{U}$). In this case all hypotheses of the fundamental principle are satisfied (cf. Theorem 2, II).

Before concluding these notes we should mention some applications of the topics treated in this volume. However the applications are manifold and too extensive to be covered in this short monograph. They pertain not only to partial differential equations but also to lacunary series, quasi-analyticity, etc. We refer the reader to Chapters VI-XIII of [23] where several applications are discussed, and many open problem suggested.

BIBLIOGRAPHICAL REMARKS AND OTHER COMMENTS

Chapter I

1 This was already known to S. Bochner in 1927 (cf. [12], Chap. VI).

2 For a thorough discussion of the role distribution played in the recent development of PDE's, the reader is referred to the beautiful monograph of F. Trèves [52].

3 One solution of this problem has been proposed by W. Słowikowski [47]. However his conditions are not formulated in terms of standard functional analysis. Another approach to this problem was worked out by V. Pták (cf. [44,45] and the references in these papers). He formulates the concept of a semiorthogonal subspace R of F. This approach uses a standard framework. In certain classes of $(\mathcal{L}\mathcal{F})$-spaces, Pták's conditions are necessary (and sufficient). For the general case, the necessity of these conditions has yet to be proved, although it is very likely that this is the case. In general, one can say that the purpose of these works is to find an abstract formulation (in terms of topological vector spaces) of Hörmander's notion of strong P-convexity [26].

4 In several concrete spaces the necessary and sufficient conditions for F to be slowly decreasing are known. Thus, e.g. Ehrenpreis [20,21] found such conditions for the space $\mathcal{E}$. In this case, F is slowly decreasing if and only if there are positive numbers a, b and c such that, for all $z \in \mathbb{C}^n$,

$$\max\{|F(z')|:|z-z'|\leq a(\log(1+|z|)+|\mathrm{Im}\ z|)\} \geq b(1+|z|)^{-c}\exp(-c|\mathrm{Im}\ z|).$$

Let us observe that here the maximum occurs instead of the minimum as in (7). To go from max to min one has to use the minimum modulus theorem [33]. Similar conditions for the Gevrey classes were given by Ch.-Ch. Chou [14].

5 Theorem 1 and its corollaries are taken from our paper [5] (for the

proof of Theorem 1, cf. also [16]).

6 The proof is based upon an idea from [22].

CHAPTER II

1 The spaces considered in this chapter were introduced by Arne Beurling in 1961 [8]. A systematic study of Beurling spaces was later published by G. Björck [9] who, in following the program of Hörmander's monograph [27], put the main emphasis on applications to partial differential equations. A theorem on regularity of solutions to elliptic partial differential equations was proved for Beurling spaces by O. John [31] (cf. also the article of E. Magenes [34]). Other problems concerning Beurling spaces are studied in our papers [6,15] and in a recent paper by G. Björck [10].

2 Propositions 1 and 2 are taken from our papers [5,6,17]. Part of the proof of Proposition 1 is based on the same idea as a theorem of B. Malgrange [36].

3 Proposition 3 and Theorem 2 appear here for the first time. It is not without interest to observe that a different construction given by L. Ehrenpreis yields Theorem 2 for the case $\mathcal{D}'(\mathbb{R}^n)$ ([23], Chap. V). However his proof is different and does not seem to generalize to Beurling spaces.

4 The proof of this fact proceeds similarly as in the classical case of $\mathcal{D}_\omega = \mathcal{D}(\mathbb{R}^n)$ [46] and it is left for the interested reader.

5 An interesting characterization in terms of approximation of Beurling test functions (and, more generally, of elements in $\mathcal{E}_\omega$) was found by G. Björck [10].

6 Proposition 4 and Theorem 3 are taken from our paper [6]. The proof of Proposition 4 follows the proof of Theorem 5.15, [23].

CHAPTER III

1 Actually, it is not necessary to assume that the distributions L_j have compact support [2,3]. Therefore the theorems of §2 can be generalized to those L satisfying conditions (42).

2 This theorem was originally proved by Ehrenpreis [23] for the case of differential operators. As was observed in [2], Ehrenpreis's proof could be extended to convolutors (cf. Theorem 1) by using the generalized form of the fundamental principle (Theorem 1, IV). The proof given below closely follows a different approach due to B. A. Taylor [50]. Both methods are based on the idea of extending certain functions in n variables to functions in n+1 variables. This can be done by considering the functions in $\mathcal{E}_B(L;\Phi)$ as different Cauchy data of a differential equation which in Ehrenpreis's proof is the heat equation. The proof given here can be interpreted as the study of the equation

$$\frac{\partial}{\partial t}\,\phi(x,t) = L * \phi(x,t)$$

in the space $E(B;\Phi)$ of all functions $\phi(x,t)$, $x \in \mathbb{R}^n$, $t \in \mathbb{R}$, such that ϕ satisfies the growth conditions on x, and, as a function of t, ϕ belongs to the space $\mathcal{E}_B$. It is clear that for any such ϕ which satisfies the above equation, $\phi(x,0) \in \mathcal{E}_B(L;\Phi)$.

3 To prove that $\mathcal{E}_B(L;\Phi)$ is an AU-space (i.e. not only that it is a weak AU-space) one should impose additional restrictions on L. For instance, when L is a differential operator, $\mathcal{E}_B(L;\Phi)$ is obviously an AU-space.

4 The content of this section simultaneously generalizes the uniqueness theorem for the heat equation and the Denjoy-Carleman theorem. For the Laplace operator this result can be traced to S. Bochner [13] and for differential operators with constant coefficients to Ehrenpreis [23]. In this section we follow [3].

CHAPTER IV

As mentioned above the fundamental principle was stated first by L. Ehrenpreis in 1960 [18]. The proof in its full generality (i.e. for systems of linear PDE's with constant coefficients) was published in his monograph [23] in 1970. In the meantime V. I. Palamodov published his version of the proof in [41]. Both proofs follow essentially the same pattern (i.e. locally extending functions from varieties; use of the Lagrange interpolation formula; proof of the vanishing of a certain cohomology group, etc.). Palamodov's proof systematically uses homological methods and the Hörmander estimates of the $\bar{\partial}$-operator [29]. A weaker version of the theorem was proved by B. Malgrange [37]. Theorem 1 of this chapter generalizes the fundamental principle for one equation to the case of distinguished polynomials (cf. Remark 3, IV). Its proof is taken from [2] and follows the method of Ehrenpreis [23]. Although the case of distinguished polynomials would not seem to be very different from the case of arbitrary polynomials, it is interesting to observe that certain "unexpected" factors appear (cf. the discriminant Δ in Theorem 1). Intuitively, the assumption on Δ says that the roots of P do not coalesce very abruptly. It seems that in order to generalize Theorem 1 further, one will have to impose a similar restriction on the geometric nature of the variety $V_p = \{z: P(z) = 0\}$. Another way of generalizing Theorem 1 is to study the case of systems. Here the problems are of an algebraic nature, and for the case of polynomials have been solved (cf. [23]). The relation of the above mentioned theorem of Malgrange to Theorem 1 can be better understood if we look at the problem from the point of view of <u>real</u> variables (i.e. the theory of distributions). First we should prove that $P(D)T = 0$ implies $\operatorname{supp} \hat{T} \subseteq V_p$ (here $\hat{T}$ is taken from the space $\hat{W}$ which is defined as the dual of $\hat{U}$); second, we should establish representation (9), IV. In the case of one variable, the geometric nature of the variety V_p is so simple that the second step follows immediately from the first one. However, it is well known (cf. [46], Chap. III, §9, §10) that the inclusion supp $\hat{T} \subseteq A$ does not imply that $\hat{T}$ is a combination of derivatives of measures with supports in A. Therefore the first step does not immediately imply representation (9). Malgrange proves essentially the first step. Ehrenpreis's method can be viewed as a way of establishing sufficient conditions for certain varieties to be regular in Whitney's sense [46]. Hence it is not unexpected that for functions P which are not polynomials, one has to impose additional restrictions as in Theorem 1, II.

2 To show that D has the form $D = c\pi \ldots$, where c is the constant defined in Lemma 4, is actually quite tedious. Let us introduce m_1 different variables $\sigma_1,\ldots,\sigma_{m_1}$, then m_2 different variables $\tau_1,\ldots,\tau_{m_2}$, etc., and consider the Vandermonde determinant of order m

$$V = \begin{vmatrix} \sigma_1^{m-1} & \sigma_1^{m-2} & \ldots & \sigma_1 & 1 \\ \sigma_2^{m-1} & \sigma_2^{m-2} & \ldots & \sigma_2 & 1 \\ \vdots & \vdots & \vdots\vdots\vdots & \vdots & \vdots \end{vmatrix} .$$

Then the value of $V = (\sigma_2-\sigma_1)(\sigma_3-\sigma_1)\ldots(\sigma_{m_1}-\sigma_1)\ldots$, does not change if we subtract the first row from the following m_1-1 rows containing σ's; then the row of τ_1 from the following rows containing τ's, etc. Let us divide V by $(\sigma_2-\sigma_1)(\sigma_3-\sigma_1)\ldots(\sigma_{m_1}-\sigma_1)(\tau_2-\tau_1)(\tau_3-\tau_1)\ldots$. This is equivalent to the division of the row of σ_2 by $\sigma_2-\sigma_1$, etc. Now we subtract the new row of σ_2 from the new rows of σ's that follow, and divide by $(\sigma_3-\sigma_2)(\sigma_4-\sigma_2)\ldots$. In the end the resulting determinant will not contain any term of the form $\sigma_j-\sigma_i$, $\tau_j-\tau_i$, ... ; and, the entries will be certain divided differences. Now we can take $\sigma_j \to s_1$, $\tau_k \to s_2$, ..., and the limit will be

$$(s_2-s_1)^{m_1m_2}\,(s_3-s_1)^{m_3m_1}\ \ldots\ .$$

Moreover, the entries in the resulting determinant are the desired quantities divided by the corresponding factorials (for example, the original row of σ_3 appears divided by 2!, etc.). The value of D is then the square of the previous determinant.

Bibliography

[1] BANG, T., Om quasi-analytiske funktioner, thesis, University of Copenhagen, 1946.

[2] BERENSTEIN, C.A., Convolution operators and related quasi-analytic classes, Ph.D. thesis, New York University, 1970.

[3] BERENSTEIN, C.A., An uniqueness theorem for convolution operators, to be published in Commun. Pure Appl. Math.

[4] BERENSTEIN, C.A., and DOSTAL, M.A., Structures analytiques uniformes dans certains espaces de distributions, C.R. Acad. Sci., 268(1969) pp. 146-49.

[5] BERENSTEIN, C.A., and DOSTAL, M.A., Topological properties of analytically uniform spaces, Trans. Amer. Math. Soc., 154(1971), pp. 493-513.

[6] BERENSTEIN, C.A., and DOSTAL, M.A., Fourier transforms of the Beurling classes $\mathcal{D}_\omega, \mathcal{E}'_\omega$, Bull. Amer. Math. Soc., 77(1971), pp. 963-967.

[7] BERENSTEIN, C.A., and DOSTAL, M.A., Some remarks on convolution equations (in preparation).

[8] BEURLING, A., Quasi-analyticity and general distributions, Lectures 4 and 5, Amer. Math. Soc. Summer Institute, Stanford, 1961 (mimeographed).

[9] BJÖRCK, G., Linear partial differential operators and generalized distributions, Arkiv f. Mat., 6(1966), pp. 351-407.

[10] BJÖRCK, G., Beurling distributions and linear partial differential equations, Atti di Convegno sulle equazioni ipoellittiche e gli spazi funzionali, Roma 1971 (to be published).

[11] BOAS, R.P., "Entire functions," Academic Press, New York, 1954.

[12] BOCHNER, S., "Lectures on Fourier integrals," Princeton University Press, Princeton, N.J., 1959.

[13] BOCHNER, S., Partial differential equation and analytic continuation, Proc. Nat. Acad. Sci., 38(1952), pp. 227-30.

[14] CHOU, Ch.-Ch., La Transformation de Fourier et l'Equation de Convolution (Thèse), Faculté des Sciences, Nice, 1969-1970.

[15] DOSTAL, M.A., An analogue of a theorem of Vladimir Bernstein and its applications to singular supports of distributions, Proc. London Math. Soc., 19(1969), pp. 553-576.

[16] DOSTAL, M.A., Espaces analytiquement uniformes, "Séminaire Pierre Lelong (Analyse), Année 1970," Lecture Notes in Math., no. 205, Springer-Verlag, 1971, pp. 113-128.

[17] DOSTAL, M.A., A complex characterization of the Schwartz space $\mathcal{D}(\Omega)$, to appear in Math. Annalen.

[18] EHRENPREIS, L., The fundamental principle for linear constant coefficient partial differential equations, "Proc. Intern. Symp. Lin. Spaces, Jerusalem, 1960," Jerusalem, 1961, pp. 161-174.

[19] EHRENPREIS, L., Analytically uniform spaces and some applications, Trans. Amer. Math. Soc., 101(1961), pp. 52-74.

[20] EHRENPREIS, L., Solutions of some problems of division III, Amer. J. Math., 76(1954), pp. 883-903.

[21] EHRENPREIS, L., Solutions of some problems of division IV, ibid., 82(1960), pp. 522-588.

[22] EHRENPREIS, L., Theory of infinite derivatives, ibid., 81(1959), pp. 799-845.

[23] EHRENPREIS, L., "Fourier analysis in several complex variables," Wiley-Interscience, New York, 1970.

[24] FLORET, K., and WŁOKA, J., "Einführung in die Theorie der lokalkonvexe Räume," Lecture Notes in Math., no. 56, Springer-Verlag, Berlin, 1968.

[25] GUNNING, R.C., AND ROSSI, H., "Analytic functions of several complex variables," Prentice-Hall, 1965.

[26] HÖRMANDER, L., On the range of convolution operators, Ann. Math., 76(1962), pp. 148-170.

[27] HÖRMANDER, L., "Linear partial differential operators," Springer-Verlag, 1963.

[28] HÖRMANDER, L., Supports and singular supports of convolutions, Acta Math., 110(1965), pp. 279-302.

[29] HÖRMANDER, L., "An introduction to complex analysis in several variables," Van Nostrand, Princeton, 1965.

[30] JOHN, F., On linear partial differential equations with analytic coefficients, Commun. Pure Appl. Math., 2(1949), pp. 209-253.

[31] JOHN, O., Sulla regolarità delle soluzioni delle equazioni lineari ellitiche nelle classi di Beurling, Boll. Un. Mat. Ital., 4(1969), pp. 183-195.

[32] KÖTHE, G., "Topologische lineare Räume," Vol. I, Die Grundlehren der math. Wissenschaften, Band 107, Springer-Verlag, Berlin, 1960.

[33] LEVIN, B.Ja., "Distribution of zeros of entire functions," Moscow, 1956; (English translation published by Amer. Math. Soc., Providence, R.I., 1964)

[34] MAGENES, E., Alcuni aspetti della teoria delle ultradistribuzioni e delle equazioni a derivate parziali, Simposia Matematica, Vol. II, Istituto Nazionale di Alta Matematica, Acad. Press, 1969, pp. 235-254.

[35] MALGRANGE, B., Existence et approximation des solutions des équations aux derivées partielles et des équations de convolution, Ann. l'Inst. Fourier, 6(1956), pp. 271-355.

[36] MALGRANGE, B., Sur la propagation de la régularité des solutions des équations à coefficients constants, Bull. Math. Soc. Sci. Math. Phys. R.P. Roumaine, 51 (1959), pp. 433-440.

[37] MALGRANGE, B., Sur les systèmes differentiels à coefficients constants,"Séminaire sur les équations aux dérivées partielles," Collège de France, 1961-62; also in "Séminaire Bourbaki," 1962-63, No. 246; also in "Les équations aux dérivées partielles," Paris, 1963, pp. 113-122.

[38] MANDELBROJT, S., "Séries adhérentes, régularisation des suites, applications," Paris, 1952.

[39] MARTINEAU, A., Sur les fonctionelles analytiques et la transformation de Fourier-Borel, J. d'Analyse Math., 9(1963), pp. 1-144.

[40] OSTROWSKI, A., Recherches sur la méthode de Graeffe et les zéros de polynômes et des séries de Laurent, Acta Mathematica, 72(1940-41), pp. 99-257.

[41] PALAMODOV, V.P., "Linear differential operators with constant coefficients," Moscow, 1967 (English translation published as vol. 168 of the series "Grundlehren der math. Wiss.," Springer-Verlag, 1970)

[42] PALEY, R.E.A.C., and WIENER, N., "Fourier transforms in the complex domain," Amer. Math. Soc., New York, 1934.

[43] PIETSCH, A., "Nukleare lokalkonvexe Räume," Akademie-Verlag, Berlin, 1965.

[44] PTÁK, V., Simultaneous extension of two functionals, Czech. Math. J., 19(1969), pp. 553-566.

[45] PTÁK, V., Extension of sequentially continuous functionals in inductive limits of Banach spaces, ibid., 20(1970), pp. 112-21.

[46] SCHWARTZ, L., "Théorie des distributions," Vol. I, II, Act. Sci. Industr. 1121-22, Hermann, Paris, 1950-51.

[47] SŁOWIKOWSKI, W., "Epimorphisms of adjoints to generalized $(\mathcal{LF})$-spaces," Aarhus Universitet, Matematisk Institut Lecture Notes, Aarhus, 1966.

[48] TÄCKLIND, S., Sur les classes quasianalytiques des solutions des équations aux dérivées partielles du type parabolique, Nova Acta Regiae Societatis Scientiarum Upsaliansis, 10(1936), pp. 1-56.

[49] TAYLOR, B.A., A seminorm topology for some $(\mathcal{DF})$-spaces of entire functions, to appear in Duke Math. J.

[50] TAYLOR, B.A., Analytically uniform spaces of infinitely differentiable functions, Commun. Pure Appl. Math., 24(1971), pp. 39-51.

[51] TAYLOR, B.A., Discrete sufficient sets for some spaces of entire functions, to appear in Trans. Amer. Math. Soc.

[52] TRÈVES, F., "Linear partial differential equations with constant coefficients," Gordon & Breach, New York, 1966.

[53] ZYGMUND, A., "Trigonometric series," Vol. I, II, 2nd ed., Cambridge, 1959.

Lecture Notes in Mathematics

Comprehensive leaflet on request

Vol. 74: A. Fröhlich, Formal Groups. IV, 140 pages. 1968. DM 16,–

Vol. 75: G. Lumer, Algèbres de fonctions et espaces de Hardy. VI, 80 pages. 1968. DM 16,–

Vol. 76: R. G. Swan, Algebraic K-Theory. IV, 262 pages. 1968. DM 18,–

Vol. 77: P.-A. Meyer, Processus de Markov: la frontière de Martin. IV, 123 pages. 1968. DM 16,–

Vol. 78: H. Herrlich, Topologische Reflexionen und Coreflexionen. XVI, 166 Seiten. 1968. DM 16,–

Vol. 79: A. Grothendieck, Catégories Cofibrées Additives et Complexe Cotangent Relatif. IV, 167 pages. 1968. DM 16,–

Vol. 80: Seminar on Triples and Categorical Homology Theory. Edited by B. Eckmann. IV, 398 pages. 1969. DM 20,–

Vol. 81: J.-P. Eckmann et M. Guenin, Méthodes Algébriques en Mécanique Statistique. VI, 131 pages. 1969. DM 16,–

Vol. 82: J. Wloka, Grundräume und verallgemeinerte Funktionen. VIII, 131 Seiten. 1969. DM 16,–

Vol. 83: O. Zariski, An Introduction to the Theory of Algebraic Surfaces. IV, 100 pages. 1969. DM 16,–

Vol. 84: H. Lüneburg, Transitive Erweiterungen endlicher Permutationsgruppen. IV, 119 Seiten. 1969. DM 16,–

Vol. 85: P. Cartier et D. Foata, Problèmes combinatoires de commutation et réarrangements. IV, 88 pages. 1969. DM 16,–

Vol. 86: Category Theory, Homology Theory and their Applications I. Edited by P. Hilton. VI, 216 pages. 1969. DM 16,–

Vol. 87: M. Tierney, Categorical Constructions in Stable Homotopy Theory. IV, 65 pages. 1969. DM 16,–

Vol. 88: Séminaire de Probabilités III. IV, 229 pages. 1969. DM 18,–

Vol. 89: Probability and Information Theory. Edited by M. Behara, K. Krickeberg and J. Wolfowitz. IV, 256 pages. 1969. DM 18,–

Vol. 90: N. P. Bhatia and O. Hajek, Local Semi-Dynamical Systems. II, 157 pages. 1969. DM 16,–

Vol. 91: N. N. Janenko, Die Zwischenschrittmethode zur Lösung mehrdimensionaler Probleme der mathematischen Physik. VIII, 194 Seiten. 1969. DM 16,80

Vol. 92: Category Theory, Homology Theory and their Applications II. Edited by P. Hilton. V, 308 pages. 1969. DM 20,–

Vol. 93: K. R. Parthasarathy, Multipliers on Locally Compact Groups. III, 54 pages. 1969. DM 16,–

Vol. 94: M. Machover and J. Hirschfeld, Lectures on Non-Standard Analysis. VI, 79 pages. 1969. DM 16,–

Vol. 95: A. S. Troelstra, Principles of Intuitionism. II, 111 pages. 1969. DM 16,–

Vol. 96: H.-B. Brinkmann und D. Puppe, Abelsche und exakte Kategorien, Korrespondenzen. V, 141 Seiten. 1969. DM 16,–

Vol. 97: S. O. Chase and M. E. Sweedler, Hopf Algebras and Galois theory. II, 133 pages. 1969. DM 16,–

Vol. 98: M. Heins, Hardy Classes on Riemann Surfaces. III, 106 pages. 1969. DM 16,–

Vol. 99: Category Theory, Homology Theory and their Applications III. Edited by P. Hilton. IV, 489 pages. 1969. DM 24,–

Vol. 100: M. Artin and B. Mazur, Etale Homotopy. II, 196 Seiten. 1969. DM 16,–

Vol. 101: G. P. Szegö et G. Treccani, Semigruppi di Trasformazioni Multivoche. VI, 177 pages. 1969. DM 16,–

Vol. 102: F. Stummel, Rand- und Eigenwertaufgaben in Sobolewschen Räumen. VIII, 386 Seiten. 1969. DM 20,–

Vol. 103: Lectures in Modern Analysis and Applications I. Edited by C. T. Taam. VII, 162 pages. 1969. DM 16,–

Vol. 104: G. H. Pimbley, Jr., Eigenfunction Branches of Nonlinear Operators and their Bifurcations. II, 128 pages. 1969. DM 16,–

Vol. 105: R. Larsen, The Multiplier Problem. VII, 284 pages. 1969. DM 18,–

Vol. 106: Reports of the Midwest Category Seminar III. Edited by S. Mac Lane. III, 247 pages. 1969. DM 16,

Vol. 107: A. Peyerimhoff, Lectures on Summability. III, 111 pages. 1969. DM 16,–

Vol. 108: Algebraic K-Theory and its Geometric Applications. Edited by R. M. F. Moss and C. B. Thomas. IV, 86 pages. 1969. DM 16,–

Vol. 109: Conference on the Numerical Solution of Differential Equations. Edited by J. Ll. Morris. VI, 275 pages. 1969. DM 18,–

Vol. 110: The Many Facets of Graph Theory. Edited by G. Chartrand and S. F. Kapoor. VIII, 290 pages. 1969. DM 18,–

Vol. 111: K. H. Mayer, Relationen zwischen charakteristischen Zahlen. III, 99 Seiten. 1969. DM 16,–

Vol. 112: Colloquium on Methods of Optimization. Edited by N. N. Moiseev. IV, 293 pages. 1970. DM 18,–

Vol. 113: R. Wille, Kongruenzklassengeometrien. III, 99 Seiten. 1970. DM 16,–

Vol. 114: H. Jacquet and R. P. Langlands, Automorphic Forms on GL (2). VII, 548 pages. 1970. DM 24,–

Vol. 115: K. H. Roggenkamp and V. Huber-Dyson, Lattices over Orders I. XIX, 290 pages. 1970. DM 18,–

Vol. 116: Séminaire Pierre Lelong (Analyse) Année 1969. IV, 195 pages. 1970. DM 16,–

Vol. 117: Y. Meyer, Nombres de Pisot, Nombres de Salem et Analyse Harmonique. 63 pages. 1970. DM 16,–

Vol. 118: Proceedings of the 15th Scandinavian Congress, Oslo 1968. Edited by K. E. Aubert and W. Ljunggren. IV, 162 pages. 1970. DM 16,–

Vol. 119: M. Raynaud, Faisceaux amples sur les schémas en groupes et les espaces homogènes. III, 219 pages. 1970. DM 16,–

Vol. 120: D. Siefkes, Büchi's Monadic Second Order Successor Arithmetic. XII, 130 Seiten. 1970. DM 16,–

Vol. 121: H. S. Bear, Lectures on Gleason Parts. III, 47 pages. 1970. DM 16,–

Vol. 122: H. Zieschang, E. Vogt und H.-D. Coldewey, Flächen und ebene diskontinuierliche Gruppen. VIII, 203 Seiten. 1970. DM 16,–

Vol. 123: A. V. Jategaonkar, Left Principal Ideal Rings. VI, 145 pages. 1970. DM 16,–

Vol. 124: Séminare de Probabilités IV. Edited by P. A. Meyer. IV, 282 pages. 1970. DM 20,–

Vol. 125: Symposium on Automatic Demonstration. V, 310 pages. 1970. DM 20,

Vol. 126: P. Schapira, Théorie des Hyperfonctions. XI, 157 pages. 1970. DM 16,–

Vol. 127: I. Stewart, Lie Algebras. IV, 97 pages. 1970. DM 16,–

Vol. 128: M. Takesaki, Tomita's Theory of Modular Hilbert Algebras and its Applications. II, 123 pages. 1970. DM 16,–

Vol. 129: K. H. Hofmann, The Duality of Compact Semigroups and C*- Bigebras. XII, 142 pages. 1970. DM 16,–

Vol. 130: F. Lorenz, Quadratische Formen über Körpern. II, 77 Seiten. 1970. DM 16,–

Vol. 131: A Borel et al., Seminar on Algebraic Groups and Related Finite Groups. VII, 321 pages. 1970. DM 22,–

Vol. 132: Symposium on Optimization. III, 348 pages. 1970. DM 22,–

Vol. 133: F. Topsøe, Topology and Measure. XIV, 79 pages. 1970. DM 16,–

Vol. 134: L. Smith, Lectures on the Eilenberg-Moore Spectral Sequence. VII, 142 pages. 1970. DM 16,–

Vol. 135: W. Stoll, Value Distribution of Holomorphic Maps into Compact Complex Manifolds. II, 267 pages. 1970. DM 18,–

Vol. 136: M. Karoubi et al., Séminaire Heidelberg-Saarbrücken-Strasbourg sur la K-Théorie. IV, 264 pages. 1970. DM 18,

Vol. 137: Reports of the Midwest Category Seminar IV. Edited by S. MacLane. III, 139 pages. 1970. DM 16,–

Vol. 138: D. Foata et M. Schützenberger, Théorie Géométrique des Polynômes Eulériens. V, 94 pages. 1970. DM 16,–

Vol. 139: A. Badrikian, Séminaire sur les Fonctions Aléatoires Linéaires et les Mesures Cylindriques. VII, 221 pages. 1970. DM 18,–

Vol. 140: Lectures in Modern Analysis and Applications II. Edited by C. T. Taam. VI, 119 pages. 1970. DM 16,–

Vol. 141: G. Jameson, Ordered Linear Spaces. XV, 194 pages. 1970. DM 16,–

Vol. 142: K. W. Roggenkamp, Lattices over Orders II. V, 388 pages. 1970. DM 22,–

Vol. 143: K. W. Gruenberg, Cohomological Topics in Group Theory. XIV, 275 pages. 1970. DM 20,–

Vol. 144: Seminar on Differential Equations and Dynamical Systems, II. Edited by J. A. Yorke. VIII, 268 pages. 1970. DM 20,–

Vol. 145: E. J. Dubuc, Kan Extensions in Enriched Category Theory. XVI, 173 pages. 1970. DM 16,–

Please turn over